INTERACTIVE
BIOCHEMISTRY

CD-ROM and Workbook **Windows™ and Macintosh®**

DEVELOPED BY:

Charles M. Grisham

University of Virginia

Saunders College Publishing

SAUNDERS COLLEGE PUBLISHING

A division of Harcourt College Publishers

Fort Worth Philadelphia San Diego New York Orlando Austin
San Antonio Toronto Montreal London Sydney Tokyo

The author wishes to express his sincere appreciation
for the generous support of *Interactive Biochemistry*
provided by the University of Virginia, the National Foundation,
and Saunders College Publishing.

Printed in the United States of America

ISBN 0-03-026154-6

To

Rosemary Jurbala Grisham
David William Grisham
Emily Ann Grisham
Andrew Charles Grisham

With all my love

— cmg

About the Author

Charles M. Grisham was born and raised in Minneapolis, Minnesota, and educated at Benilde High School. He received his B.S. in chemistry from the Illinois Institute of Technology in 1969 and his Ph.D. in chemistry from the University of Minnesota in 1973. Following a postdoctoral appointment at the Institute for Cancer Research in Philadelphia, Mr. Grisham joined the faculty of the University of Virginia, where he is professor of chemistry. He has authored numerous papers and review articles on active transport of sodium, potassium, and calcium in mammalian systems, on protein kinase C, and on the applications of NMR and EPR spectroscopy to the study of biological systems. Mr. Grisham's work has been supported by the National Institutes of Health, the National Science Foundation, the Muscular Dystrophy Association of America, the Research Corporation, the American Heart Association, and the American Chemical Society. He is a Research Career Development Awardee of the National Institutes of Health. In 1983 and 1984 he was a Visiting Scientist at the Aarhus University Institute of Physiology, Aarhus, Denmark, and in 1999 he was the Knapp Professor of Biochemistry at the University of San Diego. Mr. Grisham has taught biochemistry and physical chemistry at the University of Virginia for 25 years. He was the recipient in 1998 of an All-University Outstanding Teaching Award from the University of Virginia. He is a member of the American Society for Biochemistry and Molecular Biology.

Acknowledgments

Interactive Biochemistry has been made possible by a unique confluence of circumstances, participants, and supporters. It began with support from the University of Virginia in the form of a Teaching + Technology Initiative Fellowship. This generous grant made possible the feasibility studies that convinced me that software could enhance the teaching and learning of biochemistry. The National Science Foundation provided the largest portion of support of this project in the form of a grant from the Course and Curriculum Development Program. Saunders College Publishing also provided a substantial grant to support the completion of this project. Without all three of these generous contributions, *Interactive Biochemistry* would not have been possible.

I am deeply indebted to the many people at each of these institutions who helped in significant ways to make this project possible.

At the University of Virginia —
> Polley A. McClure, Vice President and Chief Information Officer
> Barbara Nolan, Vice Provost
> Jude Reagan, Associate Director, Teaching Resource Center
> Marva Barnett, Director, Teaching Resource Center
> John Alexander, Manager, Instructional Technology
> Michael Tuite, Director, New Media Center

At the National Science Foundation —
> Frank Settle, Program Director, Division of Undergraduate Education
> Gary Long, Program Director, Division of Undergraduate Education

At Saunders College Publishing —
> Emily Barrosse, Publisher
> John Vondeling, Vice President and Publisher
> Sandi Kiselica, Development Editor
> Pauline Mula, Product Manager
> Kent Porter-Hamann, Acquisitions Editor
> Sarah Cherry, Manager of Budgets and Technology
> Christine Benedetto, Technology Editor
> Doris Bruey, Production Manager
> Sandi Kiselica, Developemental Editor
> Nicole Weigel, Technology Assistant
> Sarah Fitz-Hugh, Project Editor

In addition, I am deeply indebted to Reginald H. Garrett, my loyal colleague and trusted friend, whose remarkable wisdom and limitless wit have sustained me through our collaboration on two editions of our textbook, *Biochemistry*, and whose support and friendship have been constant comfort to me throughout the development of *Interactive Biochemistry*.

Finally, I wish to gratefully acknowledge the following colleagues, whose hard work and exceptional creativity not only made *Interactive Biochemistry* possible, but also made it a pleasure to conceive and execute. The information about them is current as of spring 1999.

The Interactive Biochemistry *Design Team:*

Olga Chertihin, a research assistant in the University of Virginia's Department of Pharmacology, worked on the enzyme mechanism applets for *Interactive Biochemistry*. She received her master's degree in chemistry from Moscow State University in Russia. Ms. Chertihin has two children, with whom she enjoys downhill skiing and travelling.

Melinda D. Church, who edited this manual, is a free-lance writer and editor in Columbus, Ohio. She served as a writer and administrator at the University of Virginia from 1993–1999 and holds a master's degree in journalism from Indiana University. When not writing for work or pleasure, Ms. Church enjoys painting, reading, and spending time with her husband and daughter.

Basak Coruh was project manager for *Interactive Biochemistry*. A fourth-year chemistry major at the University of Virginia, she also worked on the software's Metabolic Map Database. Ms. Coruh is from Istanbul, Turkey. She is a student researcher in the U.Va. Department of Neurology and hopes to attend medical school.

Andrew E. Crowder is a second-year computer science major at the University of Virginia. Among his contributions to *Interactive Biochemistry* are his help in constructing the software's metabolic map database and his work in implementing the Java interface for it. Mr. Crowder is from London, England, and he is an avid fencer in his spare time.

Peter Hedlund, a programmer in the University of Virginia's New Media Center, did the HTML coding for the software. He received his master's in Slavic literature from U.Va. When not sitting behind a computer screen, Mr. Hedlund enjoys camping, hiking, and playing tennis and golf.

Robert Kennedy worked as a principal programmer for *Interactive Biochemistry,* and he is continuing his collaboration with the author on additional Java and virtual reality work. He is a fifth-year student in computer science at the University of Virginia. A native of Littleton, Colorado, Mr. Kennedy once had a pet rat named Clyde and is pursuing an interest in software development and graphics.

Flora E. Lackner's work on the project included creating the protein structure tutorial exercises, drawing the metabolic maps, drawing metabolites, and helping with the virtual reality programming. She received her bachelor's degree in chemistry from the University of Virginia in 1998, and is an intern with the Nuclear Regulatory Commission. Ms. Lackner is from Sterling, Virginia. She plays the piano, flute, and piccolo.

Quoc Lu, a third-year computer science major at the University of Virginia, was a Java programmer on the project. He was born in Saigon, Vietnam, and lived in Australia before settling in the United States. When he is not in the computer lab, Mr. Lu enjoys drawing, sports, and music. He intends to pursue a master's degree in computer science.

Edward K. O'Neil began collaborating with the author on the project's first interactive exercises in 1996, and he has served as a principal programmer on *Interactive Biochemistry*. He wrote various applets and the configuration management, among many other parts of the software. Mr. O'Neil is a first-year graduate student in computer science at the University of Virginia, from which he also holds a bachelor's degree in computer science.

Maleeha Qazi, a fourth-year math major and computer science minor at the University of Virginia, worked as a Java programmer and assisted with the metabolic map database. She is a native of Lahore, Pakistan. Ms. Qazi is a classical dancer and enjoys traveling and working with children. She intends to pursue a doctorate in computer science.

Will Rourk helped to develop the virtual reality portions of this software. He currently works at the University of Virginia's New Media Center. Mr. Roark earned a bachelor's degree in architecture from Virginia Polytechnic and State University in 1996.

Introduction to the
Saunders *Interactive Biochemistry* CD-ROM

The Saunders *Interactive Biochemistry* CD-ROM for Windows and Macintosh is an interactive presentation of biochemistry for college and university students.

The goals of *Interactive Biochemistry* are:
- To illustrate the structures of complex biological molecules
- To involve the user in interactive exercises that bring biochemistry to life
- To provide animated presentations of biological processes

Features of the CD-ROM

120 Java Applets
82 Chime Tutorials
8 Virtual Reality Scenes
Illustrating Biological Chemistry in a Way That Helps Students Learn!

Table of Contents and Highlights

For each topic, the relevant chapters of Garrett and Grisham, 2/e, are listed in parentheses. The related chapters and pages also are listed in boxes in the appropriate sections of this manual.

Bioenergetic Calculations (Chapter 3, 19-27) — a Java applet (a program that runs in an html page in a web browser) that helps students with calculations of free-energy changes for the reactions of metabolism.

Amino Acid Structures and Abbreviations (Chapter 4) — a Java applet that lets students test their knowledge of amino acid structures and the one-letter and three-letter abbreviations for each amino acid.

Amino Acid Titrations (Chapter 4) — a package of Java applets that illuminate the intricacies of amino acid titrations and calculations using the Henderson-Hasselbalch equation (52 Java applets in four sections).

The Peptide Bond — Virtual Reality Modeling (Chapters 5 and 6) — a virtual reality scene controlled by a Java applet that helps students to understand phi and psi angles and the allowed conformations of the peptide bond.

The Helical Wheel — A Device for Studying the Alpha Helix (Chapter 5 and 6) — a Java applet that presents peptide sequences in a "scrollable" helical wheel. Load a peptide, and then use the black arrowheads to scroll the peptide chain through the wheel, so that you can study different segments of a helical peptide.

Protein Structure Tutorial Exercises (covers most chapters of the book) — three-dimensional modeling of proteins and nucleic acids using the Chime plug-in from MDL Information Systems. The program can be executed with clickable buttons throughout the text of each tutorial to highlight and display special features of each structure. (This section includes 82 different Chime tutorials featuring all of the proteins whose structures are presented in Garrett and Grisham, 2/e.)

The Sweet Isomers (Chapter 7) — learning about sugars and their chiral and isomeric character. These applets cover four-, five-, and six-carbon sugars, including aldoses, ketoses, linear, and cyclic forms. The user can click on each chiral carbon in a structure to invert at that position and discover a new sugar. Clicking on "D" or "L" in the name below the structure inverts all the chiral centers at once. Pairs of sugars can be compared and related (18 separate Java applets).

Phospholipid Structure (Chapter 8 and 9) — a Java applet that allows the student to construct and study the structures and full names of a variety of phospholipid molecules.

Learning About Nucleosides and Nucleotides (Chapter 11) — a Java applet that compares the structures of nucleosides and nucleotides.

The Restriction Sites of Plasmids and Genes (Chapter 13) — a Java applet that allows the student to study the restriction sites of engineered plasmids.

Enzyme Kinetics (Chapter 14) — an extensive package of Java applets that help the student to understand the graphical and mathematical relationships of enzyme kinetics (12 separate Java applets covering Michaelis-Menten kinetics and Lineweaver-Burk and Hanes-Woolf plots).

Enzyme Mechanisms (Chapters 16-27) — a package of Java applets that coaches students through the individual steps of enzyme reaction mechanisms. Student are asked to identify important features of each reaction mechanism by clicking on relevant atoms and bonds. The applets respond by drawing mechanisms in response to student input (28 different enzyme mechanisms, many of them illustrating the catalytic chemistry of coenzymes).

The Coenzyme-Catalyzed Reactions of Metabolism (Chapters 18-27) — a Java applet that randomly displays 50 different reactions from metabolism and asks students to identify the coenzymes that catalyze that reaction. Students who thoroughly master this applet know the chemistry of metabolism.

A Virtual Biochemical World (Chapters 9, 17, 21, 33, and 34) — virtual reality scenes that present working models of several molecular motors and other biochemical systems and processes. Included are virtual reality scenes of actin-myosin contraction, ATP synthase, calcium-induced calcium release, flagellar rotation, membrane flippases, peptide synthesis on the ribosome, and treadmilling in tubulin.

The Metabolic Map as a Modeling Database (Chapters 18-27) — a database constructed from Donald Nicholson's well-known Metabolic Map. This Java applet package permits the student to find a particular reaction (or particular metabolites) within the metabolic map quickly and easily and presents 2-D and 3-D views and models for the enzyme and metabolites involved in that reaction. Searches of the metabolic map database permit students to learn which metabolites and reactions are associated with particular pathways. This package provides Chime models for 600 intermediary metabolites, as well as more than 100 Chime models of the enzymes of metabolism.

Charles M. Grisham

The Table of Contents

Charles M. Grisham

Windows™ PC Installation Instructions

Windows™ PC Installation Requirements

- Pentium processor or equivalent
- Windows 95 or Windows NT OS
- 32 megabytes of RAM
- Video card and monitor with 800 x 600 pixel resolution or higher
- Netscape Navigator 4.06 or higher
- Chemscape Chime (version 1.02 recommended)
- VRML Browser: CosmoPlayer 2.1

Installation Instructions

1. Insert the CD in the CD-ROM drive of your computer.

2. Double-click the "My Computer" icon on your desktop.

3. Double-click the icon for your CD-ROM drive. This will open a window like the following:

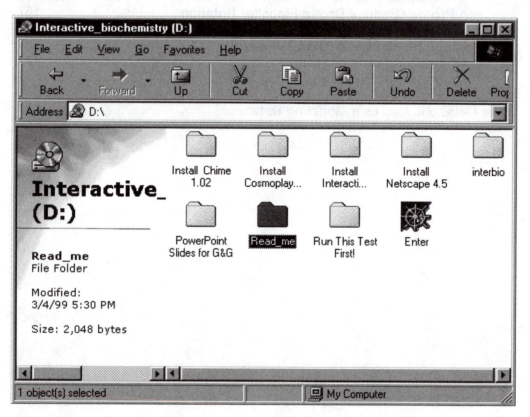

Charles M. Grisham

You must now decide which of the components to install on your computer.

4. If you are certain that the version of Netscape you are currently using is version 4.06 or higher, you may skip this step and proceed to step 5. Otherwise, uninstall any existing versions of Netscape on your computer. (Click on Start, Settings, Control Panel, Add/Remove Programs, and follow the instructions to uninstall earlier versions of Netscape.) Then, install Netscape 4.5 by opening the "Install Netscape 4.5" folder and double-clicking on the "cc32e45.exe" icon in this folder.

5. If you are certain that your current version of Netscape has Chime properly installed, you may skip this step. Otherwise, install Chime 1.02 by opening the "Install Chime 1.02" folder and double-clicking on the "ch102w32.exe" icon in this folder.

6. If you are certain that your current version of Netscape has CosmoPlayer properly installed, you may skip this step. Otherwise, install CosmoPlayer 2.1 by opening the "Install CosmoPlayer 2.1" folder and double-clicking on the "cosmo_win95nt_eng.exe" icon in this folder.

7. Install *Interactive Biochemistry* by opening the "Install Interactive Biochem" folder and double-clicking on the "Setup.exe" icon in this folder.

8. A special installation test page is provided so that you can see easily if your computer is configured properly to run *Interactive Biochemistry*. When you have completed the installations of the browser, the plug-ins, and *Interactive Biochemistry* itself, open the "Run This Test First!" folder and open the page named "Install_Test.html". If your computer is properly configured, you will see four frames, one showing an explanatory text, one showing a message that your computer is able to run Java 1.1, one showing a red cylinder in a virtual reality scene, and one showing a ribbon structure of a protein in a Chime window. If you do not see all four of these frames on the same screen, one or more of the components you will need (Netscape, Chime, or CosmoPlayer) is not properly configured. Follow the instructions to properly install the needed component(s).

Starting Interactive Biochemistry in Windows on a PC

1. Before starting *Interactive Biochemistry,* make sure the CD-ROM is in your CD-ROM drive. You cannot run *Interactive Biochemistry* without the CD-ROM in the drive.

2. To start the program, double-click the "Interactive Biochemistry" icon on your desktop. The title screen should appear, followed shortly by a Table of Contents (TOC).

Using *Interactive Biochemistry*

The Table of Contents displays all of the exercises available in *Interactive Biochemistry*. Subheadings in the TOC will display in black on the right as you sweep your cursor across the Table of Contents menu.

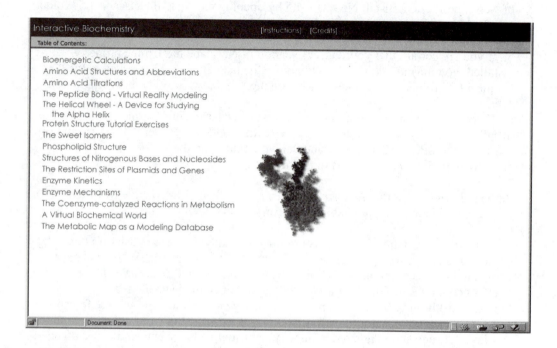

You will see a yellow navigation bar across the top of your screen. This bar shows you your pathway through the TOC. As you drop down to sub-menus in the TOC, the pathway shown in yellow reflects these movements. Note that as you sweep your mouse/cursor across the navigation bar, higher-level menus display in reversed text (white letters instead of black). Clicking your left mouse button on reversed text will return you to that menu.

Throughout the menus and exercises, help windows and instructions for running and using the exercises are available. Whenever you see a question mark (?) slightly to the right of center in the black bar at the top of your screen, you can place your cursor on the question mark (which then turns to "Information") and click to open a help/instructions window. Clicking on the main screen will close this window. The window can be enlarged or rearranged in the customary ways.

Reading the Manual and Workbook from the CD

The Manual and Workbook, including these Installation Instructions, are available on the CD (as PDF files). **To read these files and to view all the graphics,**

Charles M. Grisham

you should use the Adobe Acrobat Reader, version 4.0. You can download the Reader, version 4.0, from: http://www.adobe.com.
Follow the instructions on this site to download and install the Reader.

As shown in the figure on the preceding page, there are three clickable areas on the black bar on the *Table of Contents* screen. Clicking on *Instructions and Credits* will access this section of the Manual (with roman numeral pagination). Clicking on *Manual and Workbook* will access the section of the manual that contains the exercises (with arabic numeral pagination).

Running the Chime Tutorials

It is strongly recommended that you establish the habit of working through a Chime tutorial exactly as described in the text frame on the right side of the screen. The clickable buttons are meant to be executed in the order in which they are presented, without additional manipulations in between. However, after you have gone through the entire exercise once, you will find it even more instructive to go through the exercise a second or third time, stopping at interesting points to carry out your own manipulations of the structures, rotating, zooming, and the like. This, of course, may compromise the rest of the tutorial unless you can put the molecule back the way it was at each stage, but it will give you the chance to explore and manipulate in ways that are far more meaningful to you. Instructions for your own manipulations of the molecules are included at the end of every tutorial exercise.

Important Note about Screen Resolution

Interactive Biochemistry is designed to run optimally with 800 x 600 pixel resolution. You can run the CD and the exercises on it with 640 x 480 resolution, but the display will spill off the screen. This will make using the exercises difficult and frustrating.

Chime Operations and Mouse Actions: A Guide

Chime Action	Mac	Windows
Menu	Right	Hold down mouse button
Rotate X, Y	Left	Unmodified (click and drag)
Translate X, Y	Control-Right	Command
Rotate Z	Shift-Right	Shift-Command
Zoom	Shift-Left	Shift
Slab Plane	Control-Left	Control

Macintosh® Installation Instructions

Macintosh® Installation Requirements

- Power PC601 processor or higher
- Mac OS version 7.61 or higher
- 32 megabytes of RAM
- Video card and monitor with 800 x 600 pixel resolution or higher
- Netscape Navigator 4.0 or higher (minimum memory assigned: 16,000 kB; preferred memory assigned: 20,200 kB)
- Internet Explorer 4.01 or higher (minimum memory assigned: 10,000 kB; preferred memory assigned: 12,100 kB)
- Chemscape Chime (version 1.02 recommended)
- VRML Browser: CosmoPlayer 2.1

Installation Instructions

1. Insert the CD in the CD-ROM drive of your computer. Double-click the icon, below, that appears for the CD-ROM drive on your desktop.

This will open a window like the following:

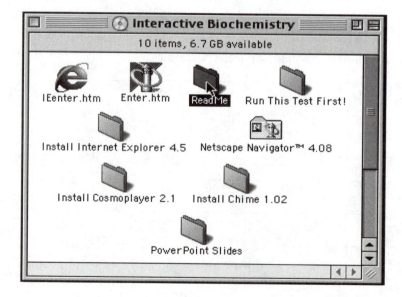

Charles M. Grisham

You must now decide which of the components to install on your computer.

2. If you are certain the version of Internet Explorer you are using is version 4.01 or higher, you may skip this step and proceed to step 3. Otherwise, install Internet Explorer 4.5 by opening the "Install Internet Explorer 4.5" folder and double-clicking on the "Internet Explorer 4.5.smi" icon in this folder.

3. If you are certain that the version of Netscape you are currently using is version 4.0 or higher, you may skip this step and proceed to step 4. Otherwise, install Netscape 4.08 by opening the "Netscape Navigator 4.08" folder and double-clicking on the "Start Here" icon in this folder.

4. If you are certain that your present version of Netscape has Chime properly installed, you may skip this step and proceed to step 5. Otherwise, install Chime 1.02 by opening the "Install Chime 1.02" folder and double-clicking on the "Chime 1.02 Installer" icon in this folder.

5. If you are certain that your present version of Netscape has CosmoPlayer 2.1 properly installed, you may skip this step and proceed to step 6. Otherwise, install CosmoPlayer 2.1 by opening the "Install CosmoPlayer 2.1" folder and double-clicking on the "Cosmo Player Install.hqx" icon in this folder.

6. The most important parameter for ensuring that *Interactive Biochemistry* will run reliably on a Mac is the amount of memory assigned to your web browser. Find the browser application icon, click on it, choose File/Get Info (Apple-I), highlight Preferred Size, and enter 20,200 for Netscape or 12,100 for Internet Explorer. Then highlight Minimum Size, and enter 16,000 for Netscape or 10,000 for Internet Explorer. If your operating system is 8.5, you will need to choose Memory from the pull-down menu to enter the appropriate values.

7. A special installation test page is provided so that you can see easily if your computer is configured properly to run *Interactive Biochemistry*. When you have completed the installations of the browsers and the plug-ins, open the "Run This Test First!" folder and open the page named "Install_Test.html" (which will open in Netscape). If your computer is properly configured, you will see four frames, one showing an explanatory text, one showing a red cylinder in a virtual reality scene, one showing a ribbon structure of a protein in a Chime window, and the fourth frame will tell you that Netscape does not display Java 1.1. If you do not see all four of these frames on the same screen, one or more of the components you will need (Netscape, Chime, or CosmoPlayer) is not properly configured. Follow the instructions to properly install the needed component(s). Next, double-click the icon named "IEinstall_test.html". The bottom left frame should tell you that Java 1.1 works in Internet Explorer. The Chime and CosmoPlayer frames will not display any content in Internet Explorer.

Macintosh Installation Notes

Installing *Interactive Biochemistry* on a Macintosh is easy, but you will need two browsers to use the software — Internet Explorer (version 4.01 or higher) and Netscape (version 4.00 or higher). You also will need two plug-ins (Chime 1.02 and CosmoPlayer 2.1) and the *Interactive Biochemistry* application itself. If you are certain that you have the correct versions of Internet Explorer, Netscape, Chime, and CosmoPlayer installed properly on your computer, you do not need to install them again. For the components you do not have installed already, simply open the installation folder and double-click the appropriate icon.

Why Do Mac Users Need Two Web Browsers?

At this writing, Java (a programming language used in *Interactive Biochemistry)* is not fully supported on Macintosh computers, and Netscape (versions 4 and below) does not support Java 1.1 well at all. Apple and Netscape plan to support Java 1.1 in Netscape 5.0, due out in the spring of 1999. Until Apple and Netscape fully support Java 1.1, Mac users will need to use Internet Explorer to view the Java applets in *Interactive Biochemistry*. Similarly, Chime and VRML are not well supported by Internet Explorer. For this reason, Mac users should use Netscape (any version) to run the parts of *Interactive Biochemistry* that use Chime and VRML.

To make this as simple as possible:

> Mac users should use Netscape for:
> • Protein Structure Tutorial Exercises
> • A Virtual Biochemical World

and use Internet Explorer 4.01 or higher for everything else on the CD.

Important Note: Because having ample memory assigned to your web browser is the surest way to run *Interactive Biochemistry* smoothly on a Mac, please double-check the memory allocation for your web browsers to make sure it matches the figures noted in step 6 on the previous page.

Starting *Interactive Biochemistry* on a Mac

1. Before starting *Interactive Biochemistry,* make sure the CD-ROM is in your CD-ROM drive. You cannot run *Interactive Biochemistry* without the CD-ROM in the drive.

2. Double-click the *Interactive Biochemistry* icon, below, on your desktop.

     Charles M. Grisham

3. Double-click the "Enter" icon, below, to start *Interactive Biochemistry* in Netscape Navigator (in order to view "A Virtual Biochemical World" and/or "Protein Structure Tutorial Exercises"),

or double-click the "IEenter" icon, below, to open *Interactive Biochemistry* in Internet Explorer (to view all the other sections listed in the Table of Contents). The title screen should appear, followed shortly by a Table of Contents (TOC).

Using *Interactive Biochemistry*

The Table of Contents displays all of the exercises available in *Interactive Biochemistry*. Subheadings in the TOC will display in black on the right as you sweep your cursor across the Table of Contents menu.

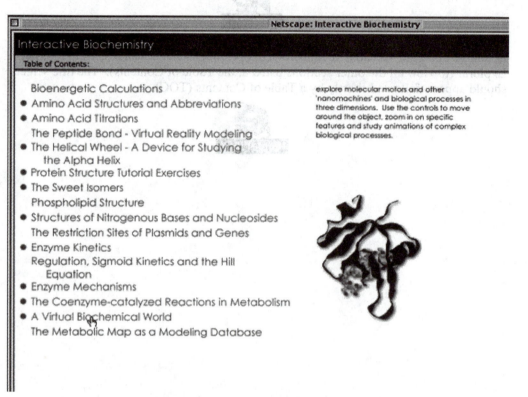

You will see a yellow navigation bar across the top of your screen. This bar shows you your pathway through the TOC. As you drop down to sub-menus in the TOC, the pathway shown in yellow reflects these movements. Note that as you sweep your mouse/cursor across the navigation bar, higher-level menus display in reversed text (white letters instead of black). Clicking on reversed text in the navigation bar will return you to that menu.

Throughout the menus and exercises, help windows and instructions for running and using the exercises are available. Whenever you see a question mark (?) slightly to the right of center in the black bar at the top of your screen, you can place your cursor on the question mark (which then turns to "Information") and click to open a help/ instructions window. Clicking on the main screen will close this window. The window can be enlarged or rearranged in the customary ways.

Reading the Manual and Workbook from the CD

The Manual and Workbook, including these Installation Instructions, are available on the CD (as PDF files). **To read these files and to view all the**

Charles M. Grisham

graphics, you should use the Adobe Acrobat Reader, version 4.0. You can download the Reader, version 4.0, from `http://www.adobe.com`. Follow the instructions on this site to download and install the Reader.

As shown in the figure on the preceding page, there are three clickable areas on the black bar on the *Table of Contents* screen. Clicking on *Instructions and Credits* will access this section of the Manual (with roman numeral pagination). Clicking on *Manual and Workbook* will access the section of the manual that contains the exercises (with arabic numeral pagination).

Running the Chime Tutorials

It is strongly recommended that you establish the habit of working through a Chime tutorial exactly as described in the text frame on the right side of the screen. The clickable buttons are meant to be executed in the order in which they are presented, without additional manipulations in between. However, after you have gone through the entire exercise once, you will find it even more instructive to go through the exercise a second or third time, stopping at interesting points to carry out your own manipulations of the structures, rotating, zooming, and the like. This, of course, may compromise the rest of the tutorial unless you can put the molecule back the way it was at each stage, but it will give you the chance to explore and manipulate in ways that are far more meaningful to you. Instructions for your own manipulations of the molecules are included at the end of every tutorial exercise.

Important Note about Screen Resolution

Interactive Biochemistry is designed to run optimally with 800 x 600 pixel resolution. You can run the CD and the exercises in it with 640 x 480 resolution, but the display will spill off the screen. This will make using the exercises difficult and frustrating.

Chime Operations and Mouse Actions: A Guide

Chime Action	*Mac*	*Windows*
Menu	Right	Hold down mouse button
Rotate X, Y	Left	Unmodified (click and drag)
Translate X, Y	Control-Right	Command
Rotate Z	Shift-Right	Shift-Command
Zoom	Shift-Left	Shift
Slab Plane	Control-Left	Control

Bioenergetic Calculations

See G&G, 2/e, p. 62, p. 614; chapts. 18-28

The free energy change, ΔG, for any reaction depends upon the nature of the reactants and products, but it is also affected by the conditions of the reaction, including temperature, pressure, pH, and the concentrations of the reactants and products. Consider a reaction between two reactants A and B that forms the products C and D:

$$A + B \rightarrow C + D$$

The free-energy change for non-standard-state concentrations is given by:

$$\Delta G = \Delta G°' + RT \ln \frac{[C][D]}{[A][B]}$$

If ΔG°' is known, then ΔG can be calculated for any combination of concentrations of A, B, C, and D, and at any temperature.

This Java applet provides an easy way to make these calculations. The calculation of non-standard-state ΔG can be performed for reactions involving one to three reactants and one to three products. When you start the applet, the display will look like this:

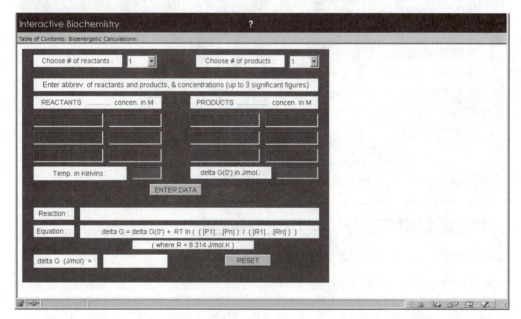

Use the drop-down boxes at the top to choose the number of reactants and products for the equation you wish to solve. Then, use the boxes in the upper middle of the display to create short names or abbreviations for the reactants and products, and their concentrations in molar (M) units. (The box where you enter reactants and products becomes lighter in appearance when you have selected the number of reactants or products above.) Finally, in the middle of the display, use the box on the left to enter the temperature (K) of the calculation and the box on the right to enter the standard-state free-energy change ($\Delta G^{\circ'}$) in J/mol. When all of these items have been entered, you may click on "ENTER DATA" to perform the calculation.

The results you should see include a reaction equation for the case you are considering, together with a calculation of the non-standard-state free energy change (shown in the lower left). In the case shown here, we consider the reaction:

Fructose-6-P + ATP $\rightarrow$ Fructose-1,6-bisP + ADP

Concentrations of these metabolites in the human red cell are:
[Fructose-6-P] = 0.000014 M (0.014 mM)
[Fructose-1,6,bisP] = 0.000031 M (0.031 mM)
[ATP] = 0.00185 M (1.85 mM)
[ADP] = 0.00014 M (0.14 mM)

The calculation at 300 K yields ΔG = -18,804 J/mol, or −18.8 kJ/mol.

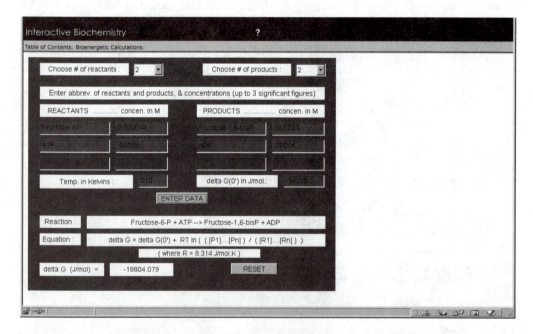

To carry out another calculation, simply click the 'RESET' button. Remember, however, that you must re-select the number of reactants and the number of products. These numbers will remain on the screen after you have clicked reset, and it may tempt you to begin trying to enter new names and concentrations. However, remember that the text entry boxes lighten when you have entered the number of reactants or products.

Charles M. Grisham

Questions

1. The standard-state free energy of hydrolysis for acetyl phosphate is $\Delta G^{\circ\prime} = -42.3$ kJ/mol:

$$\text{Acetyl-P} + H_2O \rightarrow \text{acetate} + P_i$$

Calculate the free energy change for acetyl phosphate hydrolysis in a solution of 2 mM acetate, 2 mM phosphate, and 3 nanomolar acetyl phosphate (at T = 310 K).

2. Calculate the free energy of hydrolysis of ATP in a rat liver cell at 310K, in which the ATP, ADP, and P_I concentrations are 3.4, 1.3, and 4.8 mM, respectively.

3. The citrate synthase reaction:

$$\text{Acetyl-CoA} + \text{oxaloacetate} + H_2O \rightarrow \text{CoASH} + \text{citrate}$$

Has a standard state free energy change of –31.4 kJ/mol. What is the cellular free energy change when the concentrations of acetyl-CoA and CoASH are equal, and when the concentrations of oxaloacetate and citrate are 0.3 mM and 2.2 mM, respectively.

4. Use the values given in the table below, as well as information you can obtain from any state-of-the-art biochemistry text, to work out all the cellular free energy changes for the reactions of glycolysis.

Steady-State Concentrations of Glycolytic Metabolites in Erythrocytes

Metabolite	Concentration, mM*
Glucose	5.0
Glucose-6-P	0.083
Fructose-6-P	0.014
Fructose-1,6-bisP	0.031
Dihydroxyacetone phosphate	0.14
Glyceraldehye-3-P	0.019
1,3-Bisphosphoglycerate	0.001
2,3-Bisphosphoglycerate	4.0
3-Phosphoglycerate	0.12
2-Phosphoglycerate	0.030
Phosphoenolpyruvate	0.023
Pyruvate	0.0051
Lactate	2.9
ATP	1.85
ADP	0.14
P_i	1.0

*Adapted from Minakami, S., and Yoshikawa, H., 1965. *Biochemical and Biophysics Research Communications* 18: 545.

Amino Acid Structures and Abbreviations

See G&G, 2/e
p. 82-86; chapt. 4

Introduction

The 20 common amino acids are the building blocks of proteins. Learning the structures of the amino acids and becoming conversant with their one-letter and three-letter abbreviations can be challenging at first. This exercise can help. It provides an easy way to learn the names and structures of the common amino acids. Structures of the amino acids are displayed in random order, and you are asked to identify them. Once the displayed molecule and its abbreviations are correctly identified, a new one can be displayed, and so on. This Java applet (a program that runs within a web browser) also keeps track of the user's responses, providing cumulative scoring for each session.

Name the Amino Acids

When the applet is started, a screen similar to this one appears:

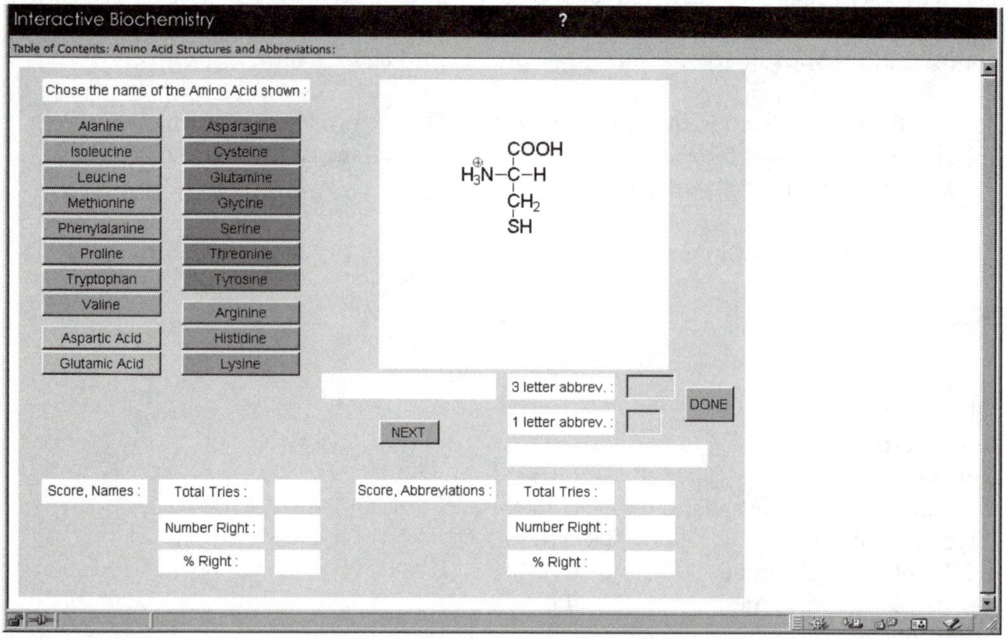

On the upper right side of the applet, you will see an image field where an amino acid is displayed. On the upper left side are groups of clickable buttons named for the 20 common amino acids. There are text areas that provide responses to your choices, and boxes in which to enter the three-letter and one-letter abbreviations of the amino acid displayed. The fields on the bottom are for scoring.

Study the displayed molecule and select and click the button bearing its name. If your selection is correct, the name you have chosen will appear in the text field under the displayed molecule. When this occurs, enter the three-letter and one-letter abbreviations of the amino acids into their respective boxes, and then click "Done". If your entries for the abbreviations are both correct, the response box below the abbreviations will answer "Correct!" You may then click the "Next" button in the middle of the screen to proceed to another case. If either of your abbreviations is incorrect, the incorrect one(s) will be deleted from the boxes, and you will be prompted to try again until you are successful. Note that the scoring area on the bottom of the screen has recorded the total number of tries (clicks) you have made, the number right, and the percentage of your total clicks that have been correct, both for the amino acid names and for the abbreviations. If you correctly identified the first displayed structure with a single click and named the abbreviations correctly, the screen will look like this:

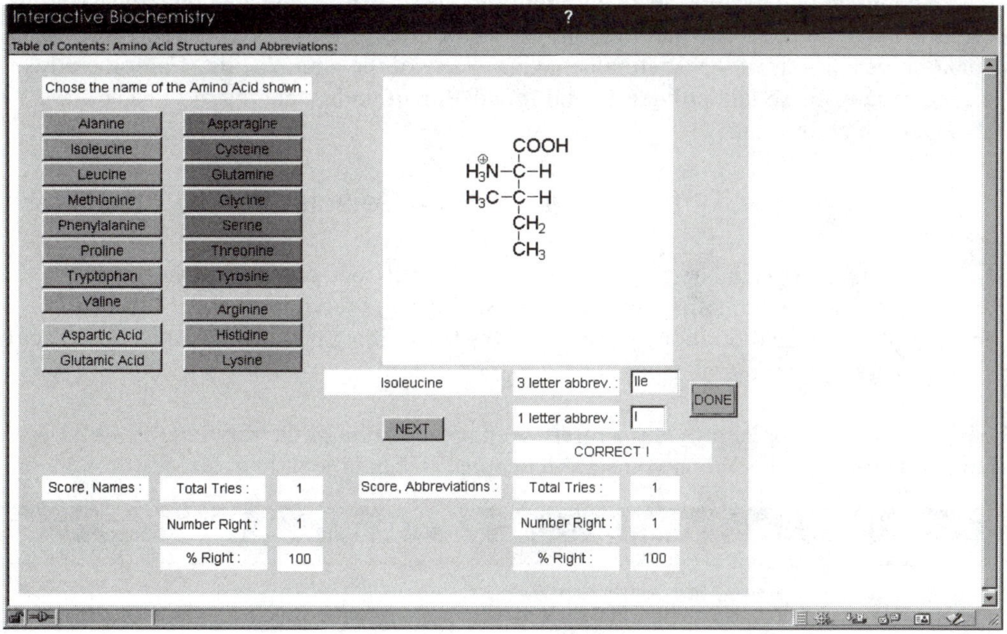

As you continue through this exercise, new molecules will be displayed every time you click "Next". The applet will keep track of your cumulative score as you proceed through the exercise.

Questions:

1. Draw the structures of all 20 of the common amino acids, noting in each case the three-letter and one-letter abbreviations.

2. Classify each of the amino acid structures you drew in question 1 as a) nonpolar, b) polar, uncharged, c) acidic, or d) basic, and describe in words the basis for your assignment.

Amino Acid Titrations

See G&G, 2/e
p. 88-92

Amino acids are polyprotic, i.e., they each possess two or three dissociable protons. The ability to accurately analyze the titration behavior of amino acids, and also the side chains of amino acids in proteins, is a prerequisite for precise and quantitative analysis of polypeptides and proteins, and their associated functions.

This section of *Interactive Biochemistry* provides a comprehensive and novel look at the titration behavior of amino acids. You can observe the titration behavior of any of the 20 common amino acids, test your ability to identify an amino acid on the basis of its titration behavior, compare the titrations of two or more amino acids, and perform quantitative analysis of the titration behavior of any of the amino acids. Careful study of these Java applets should provide a solid foundation of understanding of the acid-base chemistry of amino acids.

Observe the Titration Behavior of a Selected Amino Acid

Typical biochemistry textbooks present the titrations of at most two or three amino acids. Space limitations prevent publishers from presenting a more complete account of amino acid titrations. The student is left to imagine how the other amino acids should behave in a titration experiment.

This Java applet provides a rarely seen presentation of the titrations of all 20 common amino acids. When you start the applet, it should look like this:

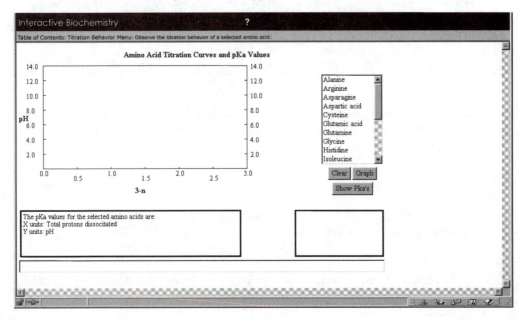

Charles M. Grisham

To view a titration, simply select (click on) an amino acid from the list in the upper right. Then click on "Graph" to display the titration curve. For example, selecting cysteine from the list, clicking on "Graph" and clicking on "Show pK$_a$s" yields the following:

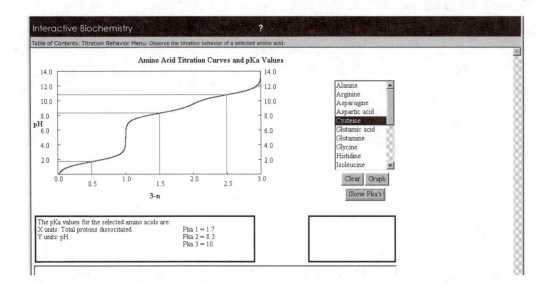

To display another titration, simply click on "Clear" and repeat the process. Select an amino acid from the list, click on "Graph" to display the titration curve, and click on "Show pK$_a$s" to see the pK$_a$ values. In this way, you can examine the titration curve of any of the 20 common amino acids, something that can't be done with a textbook.

Another useful feature: Clicking at any point in the titration will display (in the small box in the lower right corner) the concentrations of all ionic forms at that pH. For example, clicking on the curve shown in the plot above, at an x-axis value of approximately 2, yields the following display:

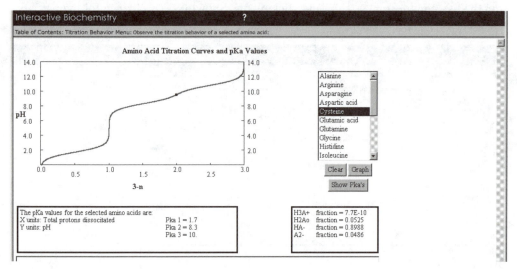

A dot appears on the curve where you have clicked. In the box, the ionic forms are represented as H_3A^+, H_2A^0, HA^-, and A^{2-}. These species represent a titration of cysteine as follows:

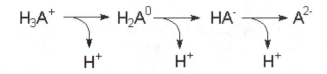

Thus, in the present case, where the curve was clicked at an x-axis value of about 2, two equivalents of protons have dissociated, so that the predominant species in the titration solution would be HA^-, and the calculations bear this out, showing that HA^- accounts for 0.8988 of the total cysteine present. Concentrations of the adjacent forms, H_2A^0 and A^{2-}, are 0.0525 and 0.0486 of the total, respectively, and the concentration of H_3A^+ is extremely small.

Clicking at other points on the titration curve will display the concentrations of each form. In this way, you can work through a titration curve, observing the concentrations of all ionic forms at any point in the curve.

Charles M. Grisham

How the Titration Curves are Calculated

This Java applet calculates each titration curve using the algorithm described by Ian Butler in his text *Ionic Equilibria - A Mathematical Approach* (1964, Addison-Wesley Publishing Co.). For the titration reaction shown on the previous page, the relevant equilibria are:

$$[H^+][H_2A^0] = K_{a1}[H_3A^+]$$

$$[H^+][HA^-] = K_{a2}[H_2A^0]$$

$$[H^+][A^{2-}] = K_{a3}[HA^-]$$

Then the mass balance on cysteine is:

$$[H_3A^+] + [H_2A^0] + [HA^-] + [A^{2-}] = 1$$

and

$$[H_2A^0] = K_{a1}[H_3A^+]/[H^+]$$

$$[HA^-] = K_{a1}K_{a2}[H_3A^+]/[H^+]^2$$

$$[A^{2-}] = K_{a1}K_{a2}K_{a3}[H_3A^+]/[H^+]^3$$

Substituting these expressions into the mass balance equation yields an equation with one unknown, as long as the K_a values are known.

Each of the curves displayed by this Java applet represents more than 2,000 calculations along the titration curve using these equations.

Identify an Amino Acid by Its Titration Behavior

A useful skill for any biochemist is the ability to identify an amino acid by its titration behavior. Being able to do this demonstrates a thorough knowledge of the acid-base chemistry of the amino acids.

This Java applet allows you to test your knowledge of amino acid titrations. When you start the applet, you will see a screen like this one:

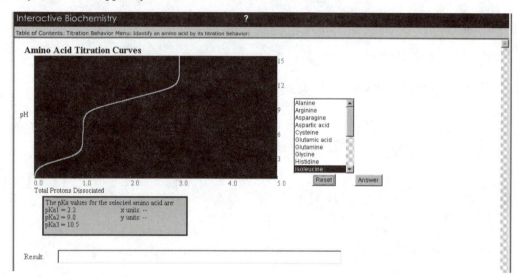

The applet presents a graph showing a randomly selected titration curve. The box below the graph displays the pK$_a$ values of the amino acid whose titration is displayed. To the right of the graph, a list box shows the names of all the amino acids. Study the titration curve in the graph, consider the pK$_a$ values shown in the box, and select the appropriate name from the box, then click on "Answer". The "Result" box at the bottom of the screen will tell you whether your choice was correct or not. If your answer was correct, you can click on "Reset" to display another (randomly selected) titration curve. If your answer was incorrect, you may select another amino acid from the list and click "Answer" again.

You will find that identifying the titration curves of amino acids with three dissociable protons and three pK$_a$ values is easier than those with two. The curves involving three dissociations are more distinctive, whereas the titration curves involving two dissociations are all quite similar to one another. To help clarify the cases with two dissociations, the table below presents the pK$_a$ values of the 20 common amino acids. The data are taken from *Biochemistry*, 2nd edition, 1999, by Reginald H. Garrett and Charles M. Grisham, Saunders College Publishing, Philadelphia.

Charles M. Grisham

pK$_a$ Values of the Common Amino Acids

Amino Acid	α-COOH pK$_a$	α-NH$_3{}^+$ pK$_a$	R group pK$_a$
Alanine	2.4	9.7	
Arginine	2.2	9.0	12.5
Asparagine	2.0	8.8	
Aspartic Acid	2.1	9.8	3.9
Cysteine	1.7	10.8	8.3
Glutamic Acid	2.2	9.7	4.3
Glutamine	2.2	9.1	
Glycine	2.3	9.6	
Histidine	1.8	9.2	6.0
Isoleucine	2.4	9.7	
Leucine	2.4	9.6	
Lysine	2.2	9.0	10.5
Methionine	2.3	9.2	
Phenylalanine	1.8	9.1	
Proline	2.1	10.6	
Serine	2.2	9.2	~13
Threonine	2.6	10.4	~13
Tryptophan	2.4	9.4	
Tyrosine	2.2	9.1	10.1
Valine	2.3	9.6	

One strategy for identifying the curves of amino acids with two dissociations would be to try the identification looking only at the graph, then consulting the pK$_a$ values in the box under the graph, and finally comparing one or both of the numbers shown in the table.

Additional note: Two of the amino acids, alanine and isoleucine, have exactly the same pK$_a$ values, 2.4 and 9.7. When you encounter a titration curve with these values, the correct answer may be **either** alanine or isoleucine, and you may have to try both to get a "correct" response from the applet. The *Interactive Biochemistry* design team programmed the applet to require one answer in some cases and the other answer in others, rather than accepting either answer whenever the 2.4, 9.7 case arose. This ensures that students will understand that both of these amino acids have the same pK$_a$ values.

Compare the Titrations of Two or More Amino Acids

One of the most fruitful exercises that can be undertaken in a study of amino acid titration behavior is the comparison of the titration curves of two or more amino acids. Comparisons permit students to analyze shapes of curves, numbers of dissociable protons, minor differences in pK$_a$ values, etc.

This Java applet makes it easy to compare two or more titration curves. When you start the applet, you will see a screen that shows an empty graph, a box below the graph, and a list box to the right of the graph. Select an amino acid from the list and click on "Graph" and the corresponding curve will be drawn on the graph. Then choose a second amino acid from the list and click on "Graph" and the second curve will be drawn. For example, if you select lysine first and histidine second, the resulting screen will look like this:

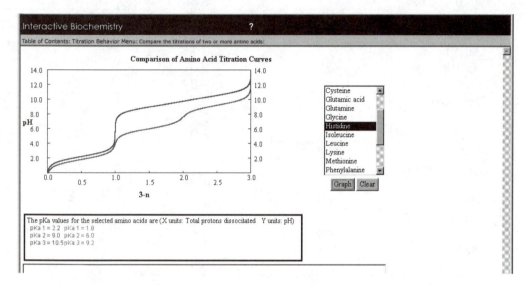

Notice that the pK_a values for lysine and histidine have been displayed in the box under the graph, in the order of their selection and graphing. Though it would be unlikely that you would want to do so, it is possible to plot up to eight titration curves on a single graph in this manner.

Analyze the Titration Behavior of a Selected Amino Acid

A thorough understanding of the titration behavior of the common amino acids requires some appreciation of the quantitative relationships that underlie these titrations. Because it is usually impractical (and beyond the scope of most undergraduate biochemistry courses) to challenge students with multi-parameter fits of amino acid titration curves, a more simplified approach is usually adopted. The Henderson-Hasselbalch equation provides an alternative in such cases. It is reasonably accurate in certain parts of the titration curve and, if its use is restricted to these regions, provides a useful acquaintance with the quantitative relationships that titration curves reflect.

When this exercise is started, it displays a list of the 20 common amino acids. Clicking on one of these yields a screen that displays a dissociation equation, together

Charles M. Grisham

with a structure of the amino acid selected. The dissociation equation in each case is reflective of the chemical state of the amino acid in each of its ionic forms.

For example, selecting arginine from the list yields a screen that looks like this:

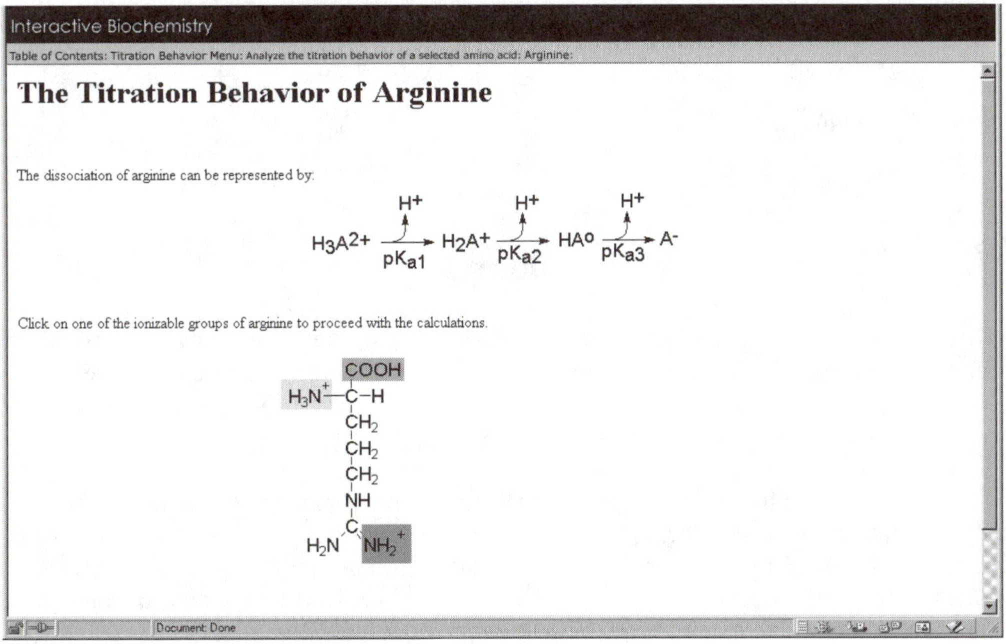

The structure shown represents the fully protonated form of arginine, which has a net charge of 2+, with two dissociable protons. Thus the equation at the top of the screen indicates that H_3A^+, the fully protonated form of the amino acid, can dissociate one proton to yield H_2A^+, and so on.

Each of the dissociable protons on the amino acid is highlighted in the structure by colored shading behind the titratable group. Each of these colored areas is clickable, and clicking on any one of them initiates a Java applet that offers calculations based on the Henderson-Hasselbalch equation. For example, clicking on the –COOH (blue) group of arginine produces a screen like the one on the next page.

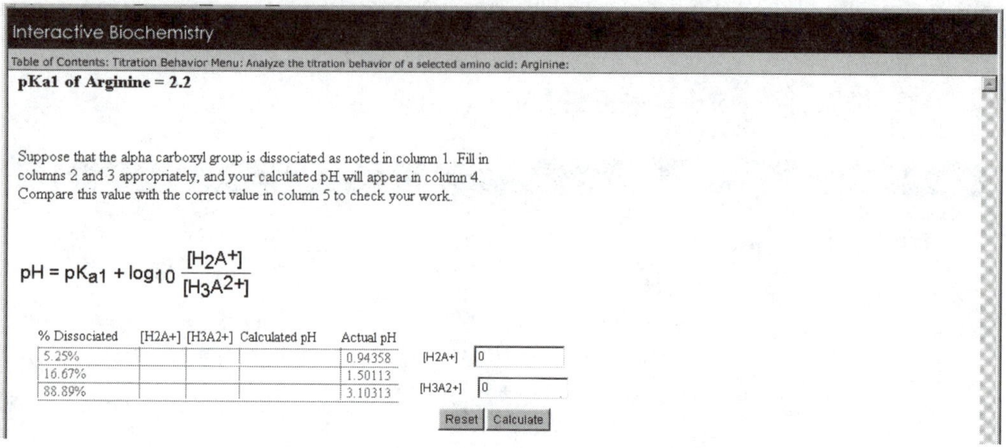

pKa1 of Arginine = 2.2

Suppose that the alpha carboxyl group is dissociated as noted in column 1. Fill in
columns 2 and 3 appropriately, and your calculated pH will appear in column 4.
Compare this value with the correct value in column 5 to check your work.

$$pH = pK_{a1} + \log_{10} \frac{[H_2A^+]}{[H_3A^{2+}]}$$

% Dissociated	[H2A+]	[H3A2+]	Calculated pH	Actual pH
5.25%				0.94358
16.67%				1.50113
88.89%				3.10313

[H2A+] 0

[H3A2+] 0

Reset Calculate

This screen shows the pK_a of the group you clicked on arginine, in this case the pK_a value
for the α-COOH, which is 2.2. Below the instructions on this screen, you will see a
Henderson-Hasselba lch equation for the dissociation of the α-COOH, and below that a
small spreadsheet and two numerical entry boxes (on the right).

To continue this exercise, consider the first number in the first column of the
spreadsheet. Suppose that the alpha carboxyl group of arginine is dissociated, as shown
in column 1. Fill in the numbers in columns 2 and 3 by entering suitable values into the
boxes on the right. In each case, simply delete the "0" (zero) from the boxes, then enter
chemically reasonable numbers into both the numerator and denominator boxes, and then
click "Calculate". The applet will enter your values in the second and third columns of
the spreadsheet and then use them to complete the Henderson-Hasselbalch calculation,
using the appropriate pK_a value — 2.2 in the present case.

In the illustration shown, the first case supposes that the alpha-carboxyl group of
arginine is 5.25% dissociated. This would mean that:

$$100 - 5.25 = 94.75\%$$

is not dissociated. The fraction dissociated is then 0.0525 and the fraction undissociated
is 0.9475. Entering 0.0525 for $[H_2A^+]$ and 0.9475 for $[H_3A^{2+}]$ and clicking "Calculate"
yields a calculated pH of 0.94358, in agreement with the "Actual pH" value shown.

Note that in this applet, the "Calculated pH" is that calculated using whatever
values are entered by the user, and the "Actual pH" is the correct calculation that the
applet has generated using the value for "% Dissociated" shown in column 1.

Completing all three calculations in the present case yields the screen shown at
the top of the following page.

Charles M. Grisham

pKa1 of Arginine = 2.2

Suppose that the alpha carboxyl group is dissociated as noted in column 1. Fill in columns 2 and 3 appropriately, and your calculated pH will appear in column 4. Compare this value with the correct value in column 5 to check your work.

$$pH = pK_{a1} + log_{10}\frac{[H_2A^+]}{[H_3A^{2+}]}$$

% Dissociated	[H2A+]	[H3A2+]	Calculated pH	Actual pH
5.25%	0.0525	0.9475	0.94358	0.94358
16.67%	0.1667	0.8333	1.50113	1.50113
88.89%	0.8889	0.1111	3.10313	3.10313

[H2A+] 0.8889

[H3A2+] 0.1111

Reset Calculate

Questions:

1. On a piece of graph paper, draw titration curves for arginine, asparagine, aspartic acid, cysteine, serine, and tryptophan. Carefully label the axes, and indicate the location and the value of each pK_a.

2. Using the first applet in this section (Observe the Titration Behavior of a Selected Amino Acid), determine the concentrations of all species in solution for each of the curves you drew in question 1 at pH values of 2.5, 4, 5.5, 7.5, and 9.

3. Consider and compare aspartic acid and glutamic acid. Explain the difference in the pK_a values for the alpha-carboxyl groups of these amino acids, on the one hand, and the two different side-chain carboxyl groups. _____
_____
_____
_____
_____

4. Consider the two amino groups of lysine, with pK_a values of 9.0 (alpha amino) and 10.5 (side chain amino). For comparison, consider ethylamine, a simple aliphatic amine with a pK_a of 10.8. Why is the pK_a of the alpha amino group so much lower than the other two values, and why is the side chain pK_a value positioned where it is?_
_____
_____
_____
_____

5. What is the pH of a dilute solution of lysine whose side-chain amino group is 70% dissociated?_____
_____

6. What is the pH of a dilute solution of proline whose alpha amino group is 60% dissociated?_____

_____

7. What is the pH of a dilute solution of cysteine whose side-chain –SH group is 40% dissociated?_____

_____

8. What are the relative concentrations of all species in a dilute solution of methionine at pH 7?_____

_____

9. What are the concentrations of all species in a 0.25 M solution of histidine at pH 5?___

_____

_____

10. Write equations for the ionic dissociations of arginine, asparagine, glycine, leucine, serine, and tryptophan.

11. The pK$_a$ values for the alpha amino groups of cysteine (10.8) and proline (10.6) are unusually high. Explain._____

_____

_____

_____

_____

_____

The Peptide Bond — Virtual Reality Modeling

See G&G, 2/e
p. 108-110, 161-163

Phi-Psi Rotations of Peptide Planes

Linus Pauling taught us that the oxygen, carbon, nitrogen, and hydrogen atoms of the peptide group, as well as the adjacent alpha-carbons, all lie in a plane. This is a necessary consequence of the partial double bond character of the peptide bond. The planarity of the peptide bond means that there are only two degrees of freedom per residue in a peptide chain. Rotation is allowed about the bond linking the alpha-carbon and the carbon of the peptide bond and also about the bond linking the nitrogen of the peptide bond and the adjacent alpha-carbon. The angle about the C(alpha)-N bond is denoted by the Greek letter phi and that about the C(alpha)-C(O) is denoted by psi. For either of these bond angles, a value of zero degrees corresponds to an orientation with the amide plane bisecting the H-C(alpha)-R(sidechain) plane and a *cis* configuration of the main chain around the rotating bond in question. The entire path of the protein backbone is known if all the phi and psi values in the protein are specified.

G.N. Ramachandran and his coworkers in Madras, India, first showed that it was convenient to plot phi values against psi values to show the distribution of allowed values in a protein or in a family of proteins. In this exercise, a pair of peptide planes joined by an alpha carbon are displayed on the top, and a Ramachandran plot is displayed on the bottom.

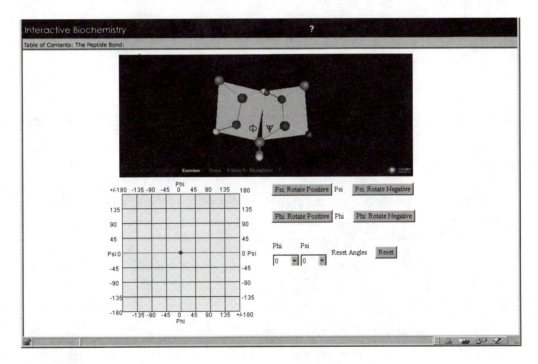

You can use the buttons on the bottom right to initiate and observe rotations about phi and psi. You also can select values of phi and psi from the choice boxes, or reset the values of phi and psi to zero with the reset button. Note when you do that the data point on the Ramachandran plot tracks and reports the phi and psi values the molecule has adopted.

You also can click and drag the data point on the Ramachandran plot to set the phi and psi angles to any value on the graph. When you do, the molecule at the top will rotate and adjust its phi and psi values to those you have selected on the graph:

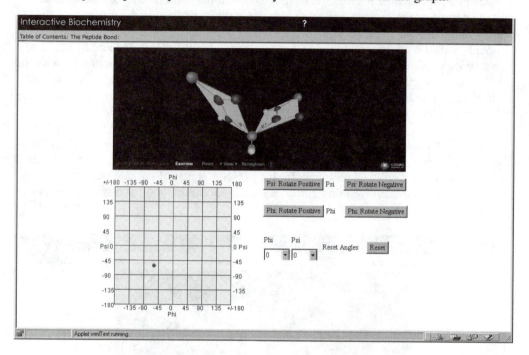

Questions:

1. Why is the combination of phi and psi values of $\phi = 0°$ and $\psi = 0°$ an unfavorable conformation?_____

_____

2. Why is the combination of phi and psi values of $\phi = 180°$ and $\psi = 0°$ an unfavorable conformation?_____

_____

3. Why is the combination of phi and psi values of $\phi = 0°$ and $\psi = 180°$ an unfavorable conformation?_____

_____

Charles M. Grisham

4. What is the combination of phi and psi values observed for a pair of peptide planes in a typical alpha helix?_____

5. What phi and psi values characterize a parallel beta sheet?_____

The Helical Wheel
— A Device for Studying the Alpha Helix

See G&G, 2/e, p. 162-168, 179-181, 318-319, 1053-1056, S-19-S-21

The ability to comprehend biological structures often depends upon your perspective. A good example is the alpha helix. This structure is based on hydrogen bonds along the backbone that stabilize the helix, but it is the side chains of the helix that interact with other structural elements and domains in a protein. Visualizing the side chains that circumscribe a helix is difficult when the helix is viewed from the side like this:

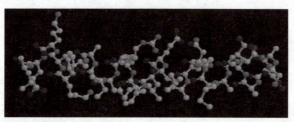

On the other hand, it often can be very useful to view the alpha helix along its axis. The **helical wheel** is a conceptual device for doing so. Shown below is a helical wheel presentation of the calmodulin-binding domain of the red blood cell protein spectrin:

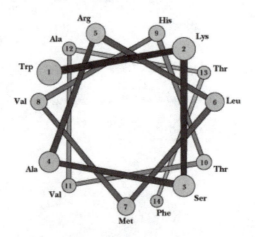

Notice that the helical wheel perspective makes clear that this helix is **amphipathic**, with one face that is hydrophobic (the lower left as shown here) and one face that is hydrophilic or polar (the upper right as shown here). This and other details of helix structure can be ascertained from the helical wheel presentation.

As useful as these displays are, they can be difficult to draw and construct. Each new amino acid in a helical wheel drawing must be exactly 100° around the wheel or circle, and it is easy to see that errors in placement of residues in such a wheel diagram

Charles M. Grisham

will accumulate. In addition, the helical wheel model is limited in scope because only 18 unique positions exist in a helical wheel diagram. If one wishes to study a longer peptide segment in a helical wheel presentation, he or she must construct two or more wheel diagrams or write more than one residue at each position. Moreover, searching the entire sequence of a protein for helical segments or putative helical segments and drawing each possible helical segment as a helical wheel can be tedious and time-consuming. This Java applet is designed to simplify and expedite such tasks.

In The Helical Wheel — a Device for Studying the Alpha Helix, students can easily display peptide segments in a helical wheel model. In addition, a long peptide sequence can be "scrolled" through the wheel diagram, so that many residues of the peptide can be studied in the wheel presentation in a short time. The applet also provides the capability to color-code the amino acids, so that clusters of polar and nonpolar amino acids and other structurally significant features can be ascertained quickly.

When the applet is started, the screen has the following appearance:

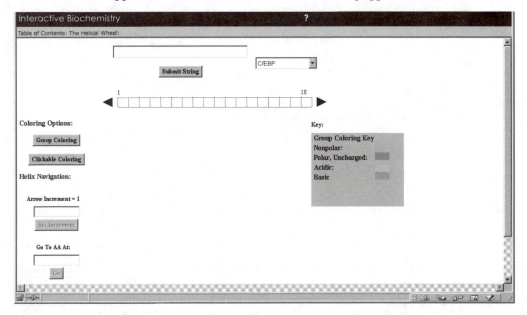

Peptides can be displayed in the wheel in two ways. First, a list box in the upper right allows the user to select one of eight peptide sequences. The peptides include segments from several bZIP (leucine zipper)-type DNA binding proteins, as well as two channel forming peptides, melittin and the magainin 2 peptide. Selecting one of these will display the peptide sequence in the window at the top of the screen. Then, clicking on "Submit String" will display the same sequence in the larger box with the arrow pointers on either side and also create a helical wheel image in the middle of the screen.

The Wheel

Several useful features are worth noting. First, the display shows 18 unique positions in the wheel, where the amino acids are represented as circles, with the N-terminal end of the sequence shown in the number 1 position, which is the largest circle in the display. As the sequence proceeds toward the C-terminal end of the peptide, the numbers increase (1 to 18) and the size of the circles decreases, so that the circle for the 18th residue is the smallest in the display.

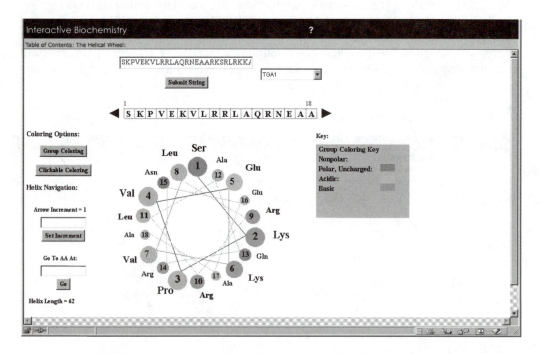

Color Coding

The circles are color-coded to indicate the character of the residue at that position. Thus, nonpolar residues are gold, polar uncharged residues are green, acidic residues are pink, and basic residues are blue. This permits the user to ascertain immediately patterns and clusters of certain types of residues anywhere in the wheel.

Two buttons on the left of the screen select Coloring Options, and they are labeled "Group Coloring" and "Clickable Coloring". Group Coloring is the default mode in which the applet begins. When Clickable Coloring is selected, the colors of all the circles are reset to gray. In this mode, the long box with the sequence shown in squares just above the circle is clickable. When you click on one of the squares containing a residue in this presentation of the sequence, the corresponding residue in the circle will turn yellow. In this way, the user can immediately see how residues in a linear sequence correspond to the residues arrayed around the helix. You may switch between Group Coloring and Clickable Coloring at any time.

Charles M. Grisham

The Scrollable Helix

One of the most useful features of the helical wheel applet is that a sequence can be "scrolled" through the helical wheel. The black triangular arrows on either side of the sequence display (above the wheel) are clickable at any time. If the sequence being examined is greater than 18 residues in length, clicking on the black arrows will scroll the peptide through the sequence box and at the same time through the helical wheel display. This is useful for examining different portions of a helical sequence or different helices in the same peptide. For example, the peptide shown on the previous page is that of TGA1, a bZIP protein that has two helical domains, one that interacts with another bZIP protein to form a coiled coil typical of leucine zipper motifs, and the other a helix that inserts into the major groove of B-DNA. The region from residues nine through 26 of this peptide is substantially basic in nature, and the positively charged residues in this region interact effectively with the phosphates of DNA. On the other hand, the C-terminus of this protein sequence possesses a classic leucine zippper motif, with Leu residues every seven residues. The scrolling arrows may be used to move rapidly between these two domains of this peptide.

One click normally advances the peptide through the displays by one residue, but there are two other modes for rapidly moving through a peptide in this applet. On the lower left of the screen is a text area labeled "Arrow Increment =" with a text entry box below it and a button reading "Set Increment". By entering a positive integer in the text entry box and clicking on Set Increment, the user can change the increment of scrolling along the sequence produced by one click of the black arrows. Alternatively, the user may go directly to a designated residue in the sequence by entering a number in the "Go to AA At" box and clicking "Go". This will revise the display to show the sequence beginning at the residue number entered in the text box.

Questions:

1. An alpha helix in flavodoxin from *Anabaena* has the sequence DDRIKSWVAELKSE. What is the character of this alpha helix?_____

    _____

2. How many leucines, seven residues apart, can you find in the leucine zipper domain of the protein GCN4?_____

3. An alpha helix in citrate synthase has the sequence LSFAAAMNGLA. What is the character of this alpha helix?_____

    _____

4. The helix described in question 3 is located on the surface of the citrate synthase monomer. What does this suggest about the normal state of this protein?_____

    _____

5. An alpha helix in calmodulin has the sequence RKMKDTDSEEEIRE. What is the character of this alpha helix?_____

    _____

6. Where would you expect to find the helix described in question 5? Consult a suitable textbook to confirm your answer._____

    _____

_____

Protein Structure Tutorial Exercises

Each of the exercises in this section of *Interactive Biochemistry* presents a protein structure in the window on the left of the screen and a tutorial script on the right. Throughout the script, which describes the structure and function of the protein shown, are clickable buttons like this:

> This enzyme is composed of 11 antiparallel beta strands. ☒
>
> These strands form an almost perfect cylinder. ☒

Each of these buttons, when clicked with the mouse (left button on a PC), manipulates and alters the displayed molecule to reveal new details of structure and function. Click the button once (and only once) to see the features of the molecule on the left as they are described in the script on the right.

Helpful Hint!

In most cases, if you click a button more than once, or if you click the buttons out of order, the displayed molecule may behave unpredictably. If this should happen, simply restart the tutorial exercise from the beginning.

You will learn that some of the buttons can be clicked more than once without affecting the actions associated with subsequent buttons. For example, a 360° rotation of a structure can be repeated any number of times without adversely affecting future portions of the tutorial.

At any point in a tutorial exercise, the displayed molecule is "live" and can be manipulated as noted here for a PC (see page xxv for Mac instructions):

- Click and hold the left mouse button to rotate the image about the x and y axes.
- Rotate about the z axis by pressing the shift key and right mouse button together.
- The image may be translated along the x and y axes by pressing control and the right mouse button.
- Press shift and the left mouse button together to zoom the image in or out.
- Clicking the right mouse button on the image gives a menu of several choices, including spinning the image and changing the appearance and color of the molecule.

Helpful Hint!

Try exploring each protein structure tutorial exercise using this two-step strategy:
1) Work through the tutorial once, clicking the buttons in order, without altering the molecule with the mouse and keyboard along the way,
2) Then repeat the tutorial exercise, stopping at one or more points to manipulate the molecule with the mouse and keyboard.

In this way, you will learn the details of structure and function described in the tutorial, and then have the chance to carry out additional explorations on your own.

Protein Secondary Structures

Alpha Helix

See G&G, 2/e
p. 162-168

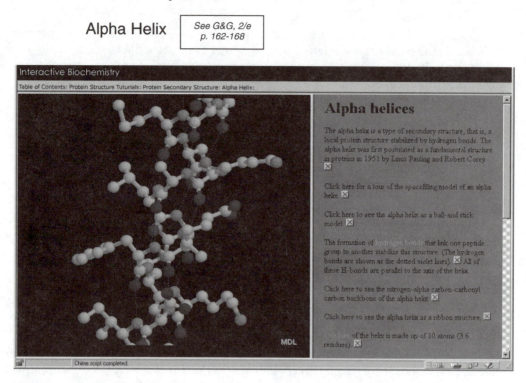

1. Who first postulated the existence of alpha helices in 1951?_____
   _____

2. One turn of an alpha helix consists of how many backbone atoms?_____

3. How many amino acid residues exist in one turn of an alpha helix?_____

4. What is the approximate diameter of an alpha helix, excluding side chains?_____

5. Each hydrogen bond closes a loop that consists of how many atoms?_____

6. How many alpha helices are found in the tertiary structure of myoglobin?_____

7. What "weak force" stabilizes the alpha helix?_____

8. In an alpha helix of 15 residues, how many hydrogen bonds are present?_____

9. What is the orientation of the hydrogen bonds in an alpha helix?_____
   _____

_____ Charles M. Grisham

Beta Sheet

See G&G, 2/e p. 168-170

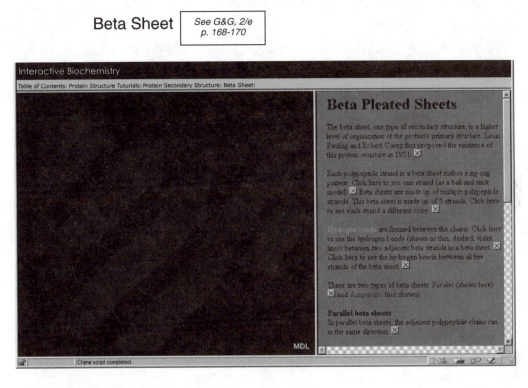

1. What forces stabilize beta sheet structures?_____

2. Who first postulated the existence of beta sheet structures?_____
   _____

3. What two types of beta sheets are found in proteins?_____
   _____

4. Which kind of beta sheet consists of peptide strands that all run in the same direction?
   _____

5. In which kind of beta sheet do alternating strands run in opposite directions?_____
   _____

6. What is the distance between residues along a single strand in a parallel beta sheet?___
   _____

7. What is the distance between residues along a single strand in an antiparallel beta
   sheet?_____

8. Where are hydrophobic residues found in parallel beta sheets?_____
   _____

9. Where are hydrophobic residues found in antiparallel beta sheets?_____
    _____

10. Based on your answers to questions 8 and 9, where would parallel beta sheets likely
    be found in globular proteins?_____

11. Similarly, based on answers to questions 8 and 9, where would antiparallel beta
    sheets likely be found in globular proteins?_____
    _____

_____

Beta Turn

See G&G, 2/e
p. 170

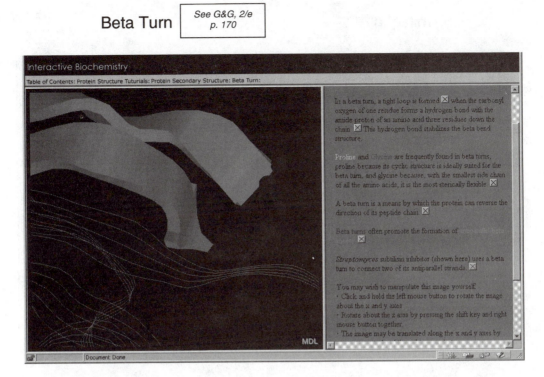

1. What are other common names for "beta turns"?_____
   _____

2. Carefully describe the atoms in a peptide chain that form a beta turn (specify kinds of atoms and relative location in the chain)._____
   _____

3. What amino acid residues are commonly found in beta turns? For each, explain WHY they are found in beta turns._____
   _____

4. Name at least one important use for beta turns in protein secondary and tertiary structure._____
   _____

Collagen

See G&G, 2/e
p. 173-179

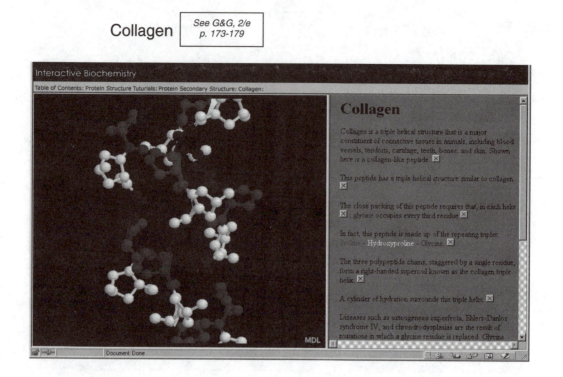

Collagen

Collagen is a triple helical structure that is a major constituent of connective tissues in animals, including blood vessels, tendons, cartilage, teeth, bones, and skin. Shown here is a collagen-like peptide. ⊠

This peptide has a triple helical structure similar to collagen. ⊠

The close packing of this peptide requires that, in each helix ⊠, glycine occupies every third residue ⊠

In fact, this peptide is made up of the repeating triplet: Proline - Hydroxyproline - Glycine. ⊠

The three polypeptide chains, staggered by a single residue, form a right-handed supercoil known as the collagen triple helix. ⊠

A cylinder of hydration surrounds this triple helix ⊠

Diseases such as osteogenesis imperfecta, Ehlers-Danlos syndrome IV, and chrondrodysplasias are the result of mutations in which a glycine residue is replaced. Glycine

MDL

Document Done

1. In what tissues of the body is collagen found?_____
   _____

2. The structure shown here is a model of collagen. What is the basic structural unit of
   collagen?_____

3. What are the dimensions of tropocollagen?_____

4. The model in this exercise has glycine at every third position, like natural collagen.
   Why is this important for the structure of collagen (name at least two reasons)?_____
   _____
   _____

5. Hydroxylated prolines and lysines are also found in large numbers in collagen. What
   is the role of hydroxyl groups in collagen structure?_____
   _____

6. Write an equation for the prolyl hydroxylase reaction. What is the vitamin that is
   required for this reaction?_____

Hydrophobic, Hydrophilic, and Amphipathic Helices

See G&G, 2/e
p. 179-181

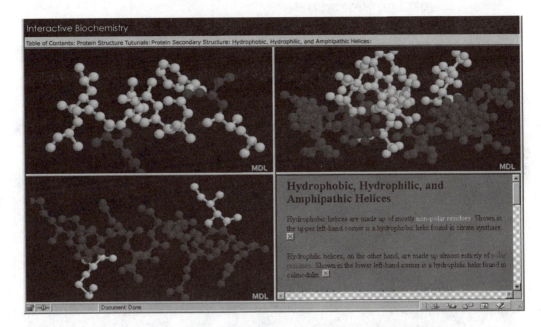

1. This exercise compares and contrasts the properties and occurrence in proteins of hydrophobic, hydrophilic, and amphipathic helices. To begin, consider the common amino acids. Is there a significant difference between polar and nonpolar amino acids in their tendency to form alpha helices?_____

_____

2. Where would you expect to find hydrophobic alpha helices in proteins? Consider the alpha helix from citrate synthase shown in this exercise. What is its character (hydrophobic, polar, or amphipathic)? Where would you think it should be located in the protein? When you worked through the exercise, is this what you found? What is the explanation?_____

3. Consider the alpha helix from calmodulin. Why is it predominantly polar?_____

_____

4. The helix from flavodoxin is amphiphilic. Why is this appropriate to its location on the surface of the enzyme?_____

_____

5. Question for further reading: Look up the sequences for at least one example each of hydrophobic, hydrophilic, and amphipathic helices. Draw helical wheel presentations of each to assess the locations of polar and nonpolar residues on each.

Representative Protein Structures

Alkaline Protease

See G&G, 2/e chapt. 6

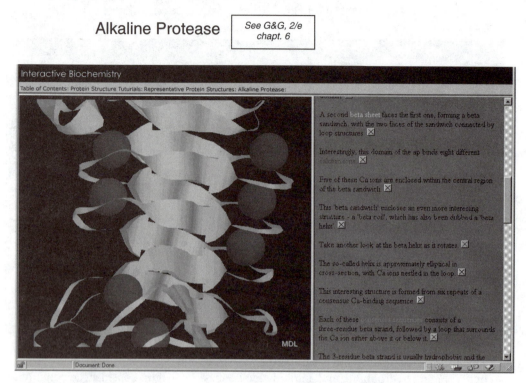

1. What two kinds of metal ions are found in the alkaline protease from *Pseudomonas aeruginosa*?_____

2. What metal ion is located in the active site of this enzyme?_____

3. What amino acid residues form ligands for this metal?_____
   _____

4. Where in the peptide chain is the "beta sandwich" or "beta helix" structure?_____
   _____

5. What metal ions are found within the beta helix structure?_____

6. What is the nine-residue consensus sequence that is repeated several times in the beta helix structure? _____
   _____

7. What are the residues that coordinate the Ca^{2+} ions in the beta helix?_____
   _____

8. What is the nature of the beta helix structure (polar, nonpolar, etc.)?_____

    _____

9. What amino acid side chains are located in the center of the beta helix, between pairs
    of Ca^{2+} ions?_____

10. What is the name of the toxin, secreted by *Bordetella pertussis*, that contains 38 of
    the same consensus repeats found in the *P. aeruginosa* alkaline protease?_____

    _____

Alpha₁-Antitrypsin

Alpha$_1$-Antitrypsin

See G&G, 2/e p. 194

Alpha$_1$-Antitrypsin

Alpha$_1$-antitrypsin is a **serine protease inhibitor** (serpin) ☒

This protein consists of several beta sheets ☒ and 9 alpha helices ☒

It prevents alveolar damage by binding to elastase, a serine protease that can break down the essential elastin found on alveolar walls. Elastase interacts with Met 358 ☒ and Ser 359 ☒, forming a complex that inactivates elastase

A deficiency of antitrypsin can lead to the destruction of alveolar walls by elastase which results in the creation of large air sacs which cannot be compressed during exhalation—a condition known as emphysema ☒

☒ lies at the N-terminal end of the reactive center loop and forms a crucial salt bridge with Lys 290 ☒ The most common deficiency occurs when Glu 342 ☒ is replaced by a lysine residue. ☒

This substitution leads to the destabilization of one of the beta sheets. This, in turn, leads to slower protein folding and

MDL

Chime script completed.

1. What is a "serpin"?_____

2. Are the beta sheets in the alpha$_1$-antitrypsin parallel, antiparallel, or mixed?_____
   _____
   _____

3. How does alpha$_1$-antitrypsin act to prevent alveolar damage from elastase? What residues of the protein are particularly important to this function?_____
   _____
   _____

4. What residues are involved in a crucial salt bridge in the protein? _____

5. How does mutation of one of these residues (question 4) result in a "deficiency" of alpha$_1$-antitrypsin, even when the protein is produced in abundance in affected individuals?_____
   _____
   _____
   _____

Charles M. Grisham

Alpha Lactalbumin

See G&G, 2/e
p. 148

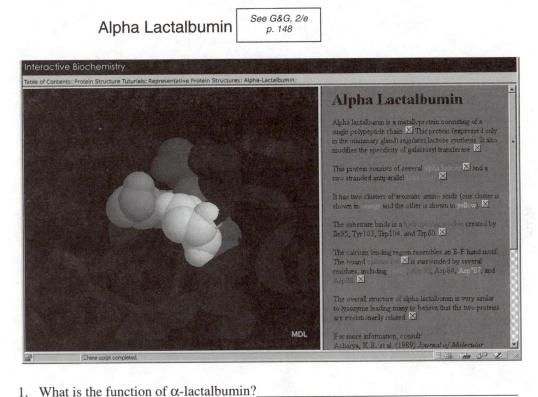

Table of Contents: Protein Structure Tutorials: Representative Protein Structures: Alpha-Lactalbumin:

Alpha Lactalbumin

Alpha lactalbumin is a metalloprotein consisting of a single polypeptide chain ☒ This protein (expressed only in the mammary gland) regulates lactose synthesis. It also modifies the specificity of galactosyl transferase ☒

This protein consists of several alpha helices ☒ and a two stranded anti-parallel ☒

It has two clusters of aromatic amino acids (one cluster is shown in orange, and the other is shown in yellow). ☒

The substrate binds in a hydrophobic pocket created by Ile95, Tyr103, Trp104, and Trp60. ☒

The calcium binding region resembles an E-F hand motif. The bound calcium ion ☒ is surrounded by several residues, including Lys79, Asp82, Asp84, Asp 87, and Asp88. ☒

The overall structure of alpha lactalbumin is very similar to lysozyme leading many to believe that the two proteins are evolutionarily related. ☒

For more information, consult:
Acharya, K.R. et al (1989) *Journal of Molecular*

1. What is the function of α-lactalbumin?_____

   _____

2. How many alpha helices are found in this structure?_____

3. How many beta sheets?_____ How many strands in each?_____

   _____

4. Which protein residues create the hydrophobic substrate binding pocket?_____

   _____

5. The calcium-binding region in α-lactalbumin is similar to what common Ca^{2+}-binding motif in proteins?_____

6. What five protein residues are involved in coordination of the Ca^{2+} ion?_____

   _____

Apolipoprotein A-1

See G&G, 2/e
p. 842

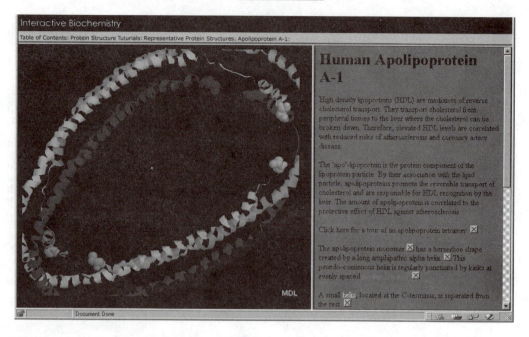

Interactive Biochemistry

Table of Contents: Protein Structure Tutorials: Representative Protein Structures: Apolipoprotein A-1:

Human Apolipoprotein A-1

High density lipoproteins (HDL) are mediators of reverse cholesterol transport. They transport cholesterol from peripheral tissues to the liver where the cholesterol can be broken down. Therefore, elevated HDL levels are correlated with reduced risks of atherosclerosis and coronary artery disease.

The 'apo'-lipoprotein is the protein component of the lipoprotein particle. By their association with the lipid particle, apolipoproteins promote the reversible transport of cholesterol and are responsible for HDL recognition by the liver. The amount of apolipoprotein is correlated to the protective effect of HDL against atherosclerosis.

Click here for a tour of an apolipoprotein tetramer ☒

The apolipoprotein monomer ☒ has a horseshoe shape created by a long amphipathic alpha helix ☒ This psuedo-continuous helix is regularly punctuated by kinks at evenly spaced ☒

A small helix located at the C-terminus, is separated from the rest ☒

MDL

Document Done

1. What is the function of human apolipoprotein A-1?_____

_____

2. What is the approximate tertiary shape of the apolipoprotein A-1 monomer?_____

_____

3. This protein is almost entirely α-helical. What is the nature or character of this helix?

_____

4. How is the nature or character of this helical structure vital to its function?_____

_____

_____

5. The amino acid sequence of this α-helical protein is punctuated by regularly spaced proline residues. Why?_____

_____

6. What effect on function would you expect for this protein if mutations converted one or more of the Pro residues into Leu residues?_____

_____

_____ Charles M. Grisham

Chymotrypsin

See G&G, 2/e
p. 118, 514-519

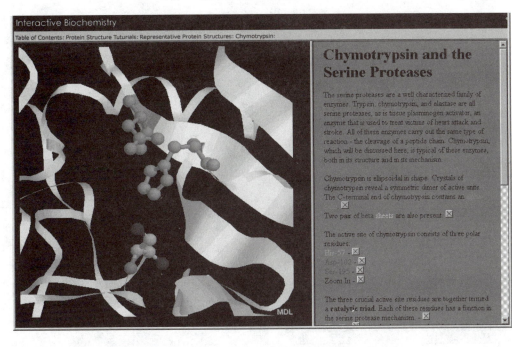

1. Chymotrypsin is a serine protease. Name three other serine proteases._____
_____

2. What is the common function of all serine proteases?_____
_____

3. What three residues are found in the catalytic triad of all serine proteases?_____
_____

4. What is the function of each of the residues in the catalytic triad?_____
_____
_____
_____

5. What is the "oxyanion hole" in the active site of chymotrypsin? What residues
   comprise it?_____
_____
_____
_____

Collagen (See page 30)

Concanavalin A See G&G, 2/e
p. 190

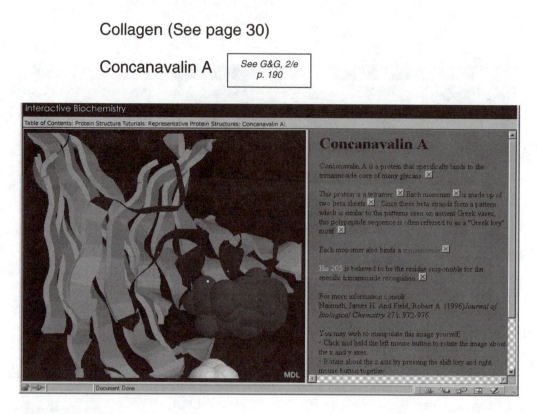

1. Concanavalin A is a protein composed of a "beta sandwich" of two beta sheets. Are these sheets parallel or antiparallel?_____

2. What is a Greek key motif in a protein?_____
_____
_____

3. Sketch below a beta sheet that possesses "Greek key topology".

4. His 205 is thought to play a role in concanavalin A, as a ligand coordinating the bound sugar moiety. Suggest a way in which histidine could play this role and draw an example of sugar coordination by a His residue._____
_____

Crambin

See G&G, 2/e, p. 191

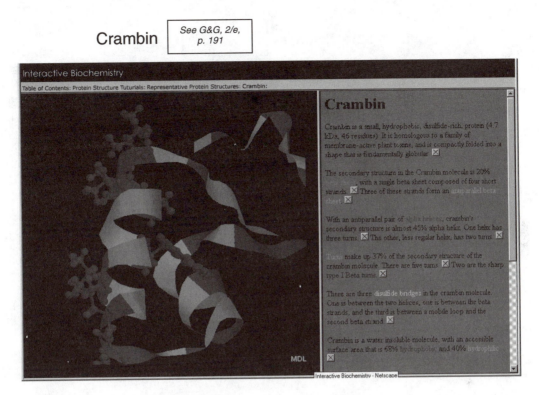

Crambin

Crambin is a small, hydrophobic, disulfide-rich, protein (4.7 kDa, 46 residues). It is homologous to a family of membrane-active plant toxins, and is compactly folded into a shape that is fundamentally globular. ☒

The secondary structure in the Crambin molecule is 20% _____ with a single beta sheet composed of four short strands. ☒ Three of these strands form an antiparallel beta sheet ☒

With an antiparallel pair of alpha helices, crambin's secondary structure is almost 45% alpha helix. One helix has three turns. ☒ The other, less regular helix, has two turns. ☒

Turns make up 37% of the secondary structure of the crambin molecule. There are five turns ☒ Two are the sharp type I Beta turns. ☒

There are three disulfide bridges in the crambin molecule. One is between the two helices, one is between the beta strands, and the third is between a mobile loop and the second beta strand. ☒

Crambin is a water insoluble molecule, with an accessible surface area that is 68% hydrophobic and 40% hydrophilic ☒

MDL

Interactive Biochemistry - Netscape

1. How many alpha helices are present in the crambin molecule?_____

2. How many strands are found in the beta sheet in crambin?_____

3. If you were to number the strands of the beta sheet beginning at the N-terminus of the protein, between which strands of the beta sheet are the alpha helices located?_____
   _____

4. How many beta turns are contained in the structure?_____

5. Where is each of the beta turns located with respect to the strands of the beta sheet?__
   _____
   _____
   _____

6. Do significant numbers of hydrophobic residues lie on the surface of this protein? ___
   _____

7. Are any of the alpha helices of crambin amphiphilic?_____

8. What do the answers to the two previous questions tell you about the behavior of crambin in aqueous solution?_____
   _____

Cytochrome Oxidase

See G&G, 2/e
p. 688-691

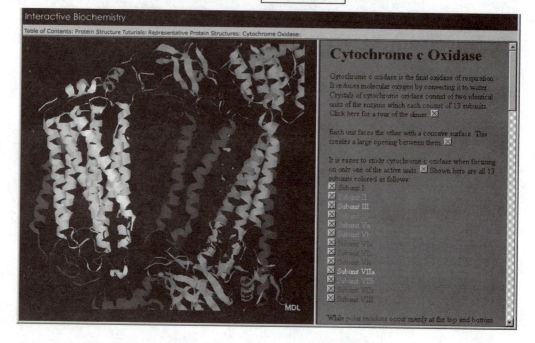

Cytochrome c Oxidase

Cytochrome c oxidase is the final oxidase of respiration. It reduces molecular oxygen by converting it to water. Crystals of cytochrome oxidase consist of two identical units of the enzyme which each consist of 13 subunits. Click here for a tour of the dimer. ☒

Each unit faces the other with a concave surface. This creates a large opening between them. ☒

It is easier to study cytochrome c oxidase when focusing on only one of the active units ☒ Shown here are all 13 subunits colored as follows:
☒ Subunit I
☒ Subunit II
☒ Subunit III
☒
☒ Subunit Va
☒ Subunit Vb
☒ Subunit VIa
☒ Subunit VIc
☒ Subunit VIIa
☒ Subunit VIIb
☒ Subunit VIIc
☒ Subunit VIII

While polar residues occur mainly at the top and bottom

MDL

1. What is the function of cytochrome c oxidase?_____
   _____

2. How many subunits are contained in a monomer of cytochrome c oxidase?_____

3. Where are polar residues concentrated on the surface of this structure? Where are nonpolar residues concentrated? What does this imply about the orientation of this protein complex in the mitochondrial membrane?_____
   _____
   _____
   _____
   _____

4. On which subunit is the heme a binding site located?_____

5. Which subunit contains a beta barrel, located outside the membrane, that contains the copper binding site of the complex?_____

6. Which subunit is dumbbell shaped, with a single transmembrane alpha helix and globular, extramembranous domains on either end?_____

7. Which subunit contains a tetrahedral zinc binding site at one end of a beta barrel?____
   _____

_____ Charles M. Grisham

8. One of these subunits contains two disulfide bonds and appears to be important in stabilizing the dimer form of this enzyme complex. Which is it?_____

9. Which amino acids play a role in the binding of the substrate, cytochrome c, to the oxidase?_____

10. What is the hypothesis for electron transport that arises from the structure of this 13-mer aggregate?_____

11. Referring to the hypothesis mentioned in question 10, which amino acid residues have been suggested to form a network for electron transfer through this complex?___

_____

_____

12. Which subunits in the cytochrome c oxidase complex do NOT contain transmembrane helical segments?_____

_____

Ferredoxin

See G&G, 2/e
p. 719-722

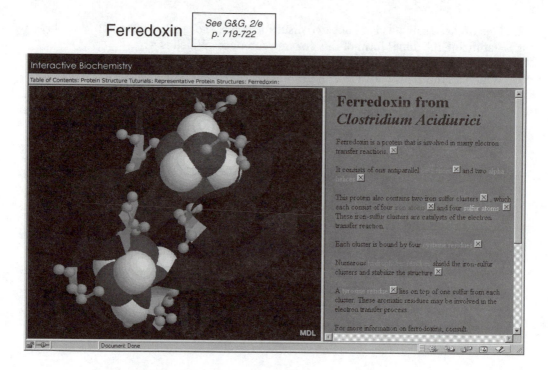

Ferredoxin from *Clostridium Acidiurici*

Ferredoxin is a protein that is involved in many electron transfer reactions. ☒

It consists of one antiparallel ☒ and two alpha helices ☒

This protein also contains two iron sulfur clusters ☒ , which each consist of four iron atoms ☒ and four sulfur atoms ☒ These iron-sulfur clusters are catalysts of the electron transfer reaction.

Each cluster is bound by four cysteine residues ☒

Numerous ☒ shield the iron-sulfur clusters and stabilize the structure ☒

A tyrosine residue ☒ lies on top of one sulfur from each cluster. These aromatic residues may be involved in the electron transfer process.

For more information on ferro-doxins, consult.

MDL

Document Done

1. This is a very simple protein structure, consisting of an antiparallel beta sheet and two alpha helices. The stability of this structure likely derives from two sources: its iron-sulfur centers and hydrophobic effects. Explain the latter effect with respect to the details of the structure._____

_____

2. Ferredoxin contains two iron-sulfur centers. Each contains four irons and four sulfurs and is coordinated by four cysteine residues from the protein. Draw the structure of the this cluster, including the cysteine groups.

3. What are the possible oxidation states of iron in iron-sulfur centers?_____

_____

4. What kinds of electron transfer processes (one-electron or two-electron) are possible for iron-sulfur centers?_____

_____

_____ Charles M. Grisham

Flavodoxin

See G&G, 2/e
p. 180, 187

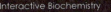

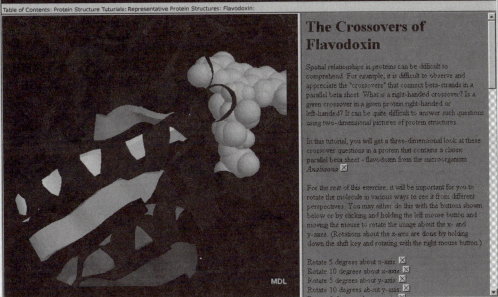

The Crossovers of Flavodoxin

Spatial relationships in proteins can be difficult to comprehend. For example, it is difficult to observe and appreciate the "crossovers" that connect beta-strands in a parallel beta sheet. What is a right-handed crossover? Is a given crossover in a given protein right-handed or left-handed? It can be quite difficult to answer such questions using two-dimensional pictures of protein structures.

In this tutorial, you will get a three-dimensional look at these crossover questions in a protein that contains a classic parallel beta sheet - flavodoxin from the microorganism *Anabaena* ☒

For the rest of this exercise, it will be important for you to rotate the molecule in various ways to see it from different perspectives. You may either do this with the buttons shown below or by clicking and holding the left mouse button and moving the mouse to rotate the image about the x- and y-axes. (Rotations about the z-axis are done by holding down the shift key and rotating with the right mouse button.)

Rotate 5 degrees about x-axis: ☒
Rotate 10 degrees about x-axis: ☒
Rotate 5 degrees about y-axis: ☒
Rotate 10 degrees about y-axis: ☒

1. Flavodoxin proteins are found in many organisms. From what organism was this particular flavodoxin obtained?_____

2. Flavodoxin contains a parallel beta sheet at its core. What is the name given to the protein segments that connect sequential strands of the parallel beta sheet?_____

3. What coenzyme is tightly bound to the protein?_____

4. The first beta strand of the protein sequence is which strand of the parallel beta sheet?

_____

5. Is the first crossover in the beta sheet right-handed or left-handed?_____

6. Which two strands of the beta sheet comprise the second crossover in the sheet?_____

_____

7. What is the handedness of the second, third, and fourth crossovers of flavodoxin?____

_____

_____

Green Fluorescent Protein

See G&G, 2/e
p. 92

Green Fluorescent Protein

The jellyfish (*Aequorea Victoria*) makes green fluorescent protein (GFP). When the jellyfish is shaken or attacked, its calcium-ion levels are altered and its aequorin then generates an excited state-which produces a blue light. This light is transferred to GFP which emits a bright green flash in response. This defensive action presumably is intended to blind, or at least startle, the jellyfish's attacker. Click here for a 360 degree tour of the enzyme. ☒

This enzyme is composed of 11 antiparallel ☒ These strands form an almost perfect cylinder ☒

The N terminus is made up of 3 antiparallel strands then a central helix (the middle of which is not shown) followed finally by another 3 antiparallel strands ☒ The peptide then crosses underneath to form the second half of the barrel ☒

The remaining antiparallel beta strands form a 3-stranded Greek key motif ☒

One short helix caps the bottom of the barrel ☒ while three short helices cap the top ☒

The p-hydroxybenzlidene- imidazolidinone chromophore is derived from a sequence that is in the middle of the single helix which lies inside the beta barrel. The residues that make up that sequence are: Ser65, Tyr66, and Gly67 ☒ Buried

MDL

1. What organism produces the green fluorescent protein?_____

2. For what purpose does it make this protein?_____

3. Green fluorescent protein captures light made by another protein and emits it at a lower energy (the phenomenon of fluorescence). What protein actually creates the light for GFP?_____

4. How many beta strands form the cylinder of green fluorescent protein?_____

5. Are these strands parallel or antiparallel?_____

6. What structures form caps at the top and at the bottom of the barrel?_____
_____

7. Name the residues that react in an autocatalytic fashion to form the chromophore of GFP._____

8. Which amino acid side chains contact (and presumably stabilize) the chromophore?__
_____

9. How is GFP used in modern biochemistry laboratories?_____
_____

Hemoglobin

See G&G, 2/e
p. 480-493

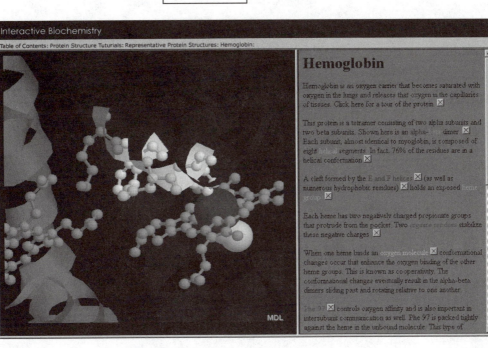

1. Hemoglobin is a tetramer consisting of two alpha and two beta subunits. How much of the tetrameric structure is shown in this tutorial?_____

2. What percentage of the residues in hemoglobin are alpha helical?_____

3. Which two helices on a hemoglobin monomer cradle the heme group?_____

4. What protein residues stabilize the negative charges on propionate residues on each heme group?_____

5. Where is Phe97 located in the Hb monomer and what is its role?_____
_____

6. Why does the iron atom normally lie above the heme plane?_____
_____

7. Which two residues on the F helix change their conformation upon O_2 binding to iron on the heme, and what does this change have to do with positive cooperativity?_____
_____

HIV-1 Protease

See G&G, 2/e p. 522-525

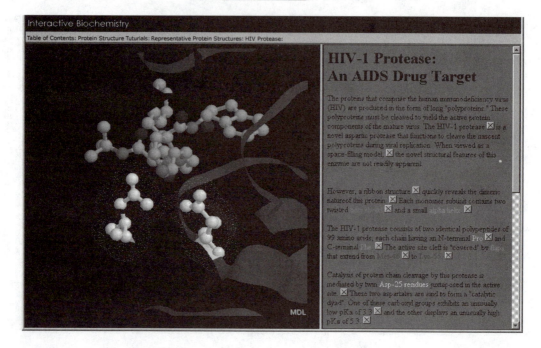

Table of Contents: Protein Structure Tutorials: Representative Protein Structures: HIV Protease:

Interactive Biochemistry

HIV-1 Protease: An AIDS Drug Target

The proteins that comprise the human immunodeficiency virus (HIV) are produced in the form of long "polyproteins." These polyproteins must be cleaved to yield the active protein components of the mature virus. The HIV-1 protease ⊠ is a novel aspartic protease that functions to cleave the nascent polyproteins during viral replication. When viewed as a space-filling model, ⊠ the novel structural features of this enzyme are not readily apparent.

However, a ribbon structure ⊠ quickly reveals the dimeric nature of this protein. ⊠ Each monomer subunit contains two twisted ⊠ and a small alpha helix. ⊠

The HIV-1 protease consists of two identical polypeptides of 99 amino acids, each chain having an N-terminal Pro ⊠ and C-terminal Phe. ⊠ The active site cleft is "covered" by flaps that extend from Met-46 ⊠ to Tyr-59. ⊠

Catalysis of protein chain cleavage by this protease is mediated by twin Asp-25 residues juxtaposed in the active site. ⊠ These two aspartates are said to form a "catalytic dyad". One of these carboxyl groups exhibits an unusually low pKa of 3.3 ⊠ and the other displays an unusually high pKa of 5.3. ⊠

1. What is the biological function of the HIV-1 protease?_____
_____

2. How many alpha helices and how many beta sheets are present in each monomer of the HIV-1 protease dimer?_____

3. What are the N-terminal and C-terminal residues of the protease sequence and where are they located?_____

4. Which residues constitute the "flaps" of the active site?_____

5. Which two residues of the dimer form the "catalytic dyad" of the active site?_____
_____

6. What are the pK_a values of these two side chains?_____

7. What drug (by which manufacturer) is shown in the active site of the protease here?__
_____

8. What two properties must a drug possess to function as a suitable orally delivered HIV treatment?_____
_____

Charles M. Grisham

Hydrophobic, Hydrophilic, and Amphipathic Helices
(See page 31)

Immunoglobulin G

See G&G, 2/e
p. 204-205

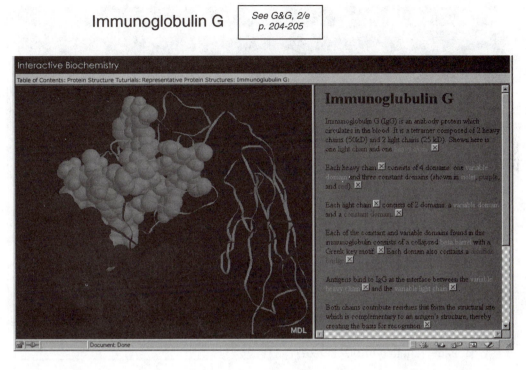

Interactive Biochemistry

Table of Contents: Protein Structure Tutorials: Representative Protein Structures: Immunoglubulin G:

Immunoglubulin G

Immunoglobulin G (IgG) is an antibody protein which circulates in the blood. It is a tetramer composed of 2 heavy chains (50kD) and 2 light chains (25 kD). Shown here is one light chain and one ☒

Each heavy chain ☒ consists of 4 domains: one variable domain and three constant domains (shown in moss, purple, and red). ☒

Each light chain ☒ consists of 2 domains: a variable domain and a constant domain. ☒

Each of the constant and variable domains found in this immunoglobulin consists of a collapsed beta barrel with a Greek key motif. ☒ Each domain also contains a disulfide bridge. ☒

Antigens bind to IgG at the interface between the variable heavy chain ☒ and the variable light chain ☒

Both chains contribute residues that form the structural site which is complementary to an antigen's structure, thereby creating the basis for recognition. ☒

MDL

Document Done

1. What is the function of immunoglobulin G (IGG)?_____
   _____

2. Explain and describe the variable domains and constant domains of the heavy and light chains of IGG._____
   _____
   _____

3. What is the fundamental protein domain that lies at the core of each of the variable and constant domains of IGG?_____

4. Where do antibodies bind to IGG?_____
   _____

5. Find and describe the location of at least four tight turns (beta turns) in this structure.
   _____

6. Question for further reading: Draw a diagram that describes the location of all of the disulfide bridges in the tetrameric IGG._____
   _____

Lysozyme

See G&G, 2/e
p. 148

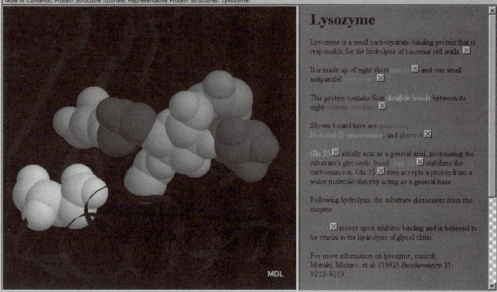

Interactive Biochemistry

Table of Contents: Protein Structure Tutorials: Representative Protein Structures: Lysozyme:

Lysozyme

Lysozyme is a small carbohydrate-binding protein that is responsible for the hydrolysis of bacterial cell walls ⊠

It is made up of eight short helices ⊠ and one small antiparallel ⊠

This protein contains four disulfide bonds between its eight cysteine residues ⊠

Shown bound here are galactose, N-acetyl-D-glucosamine, and glycerol ⊠

Glu 35 ⊠ initially acts as a general acid, protonating the substrate's glycosidic bond. Asp 53 ⊠ stabilizes the carbonium ion. Glu 35 ⊠ then accepts a proton from a water molecule-thereby acting as a general base.

Following hydrolysis, the substrate dissociates from the enzyme

⊠ moves upon inhibitor binding and is believed to be crucial in the hydrolysis of glycol chitin.

For more information on lysozyme, consult:
Muraki, Michiro, et al. (1992) *Biochemistry* 31: 9212-9219

MDL

1. What is the function of lysozyme?_____

2. How many alpha helices and beta sheets can you identify in the protein structure?____
_____

3. How many disulfide bonds are there? Where are they with respect to the alpha helices and beta sheets in the sequence of the protein?_____
_____

4. What three ligands are bound to the protein in this structure?_____
_____

Draw their structures below.

5. What are the functions of Glu35, Asp53, and Tyr63 in lysozyme?_____
_____
_____
_____

Charles M. Grisham

Myoglobin

See G&G, 2/e
p. 480-493

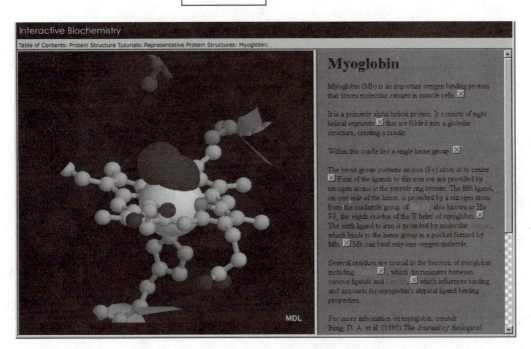

Myoglobin

Myoglobin (Mb) is an important oxygen binding protein that stores molecular oxygen in muscle cells. ☒

It is a primarily alpha helical protein. It consists of eight helical segments ☒ that are folded into a globular structure, creating a cradle.

Within this cradle lies a single heme group. ☒

The heme group contains an iron (Fe) atom at its center ☒ Four of the ligands to this iron ion are provided by nitrogen atoms in the pyrrole ring system. The fifth ligand, on one side of the heme, is provided by a nitrogen atom from the imidazole group of ☒☒, also known as His F8, the eighth residue of the 'F helix' of myoglobin. ☒ The sixth ligand to iron is provided by molecular oxygen, which binds to the heme group in a pocket formed by Mb. ☒ Mb can bind only one oxygen molecule.

Several residues are crucial to the function of myoglobin including ☒☒, which discriminates between various ligands and ☒☒, ☒ which influences binding and accounts for myoglobin's atypical ligand binding properties.

For more information on myoglobin, consult Bisig, D. A. et al. (1995) *The Journal of Biological*

MDL

1. How many alpha helices can you identify in the myoglobin structure?_____

2. What percentage of the amino acids in myoglobin (Mb) are alpha helical?_____

3. What is the usual oxidation state of Fe in the heme group of Mb?_____

4. Identify the six ligands of Fe in the heme of Mb._____
   _____

5. What is the nature of the conformation change that occurs upon oxygen binding by Mb?_____

6. How far does the Fe atom move during this conformation change?_____

7. What is the function of Val68 in Mb?_____
   _____

8. What is the function of Leu29 in Mb?_____
   _____

Myohemerythrin

See G&G, 2/e p. 186

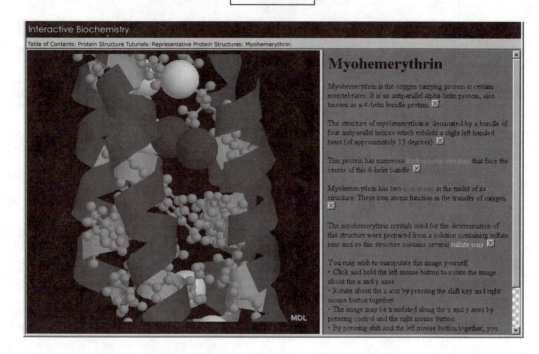

Myohemerythrin

Myohemerythrin is the oxygen carrying protein in certain invertebrates. It is an antiparallel alpha-helix protein, also known as a 4-helix bundle protein. ☒

The structure of myohemerythrin is dominated by a bundle of four antiparallel helices which exhibits a slight left handed twist (of approximately 15 degrees). ☒

This protein has numerous hydrophobic residues that face the center of this 4-helix bundle. ☒

Myohemerythrin has two iron atoms in the midst of its structure. These iron atoms function in the transfer of oxygen. ☒

The myohemerythrin crystals used for the determination of this structure were prepared from a solution containing sulfate ions and so this structure contains several sulfate ions. ☒

You may wish to manipulate this image yourself.
• Click and hold the left mouse button to rotate the image about the x and y axes
• Rotate about the z axis by pressing the shift key and right mouse button together
• The image may be translated along the x and y axes by pressing control and the right mouse button
• By pressing shift and the left mouse button together, you

1. Describe the structure of myohemerythrin._____

_____

2. This protein is water-soluble, yet is composed largely of hydrophobic amino acids.
   Explain._____

_____

3. How many Fe atoms are contained in this structure and where are they located?_____

_____

4. What do you think is the primary driving force for the folding of myohemerythrin?___

_____

_____

Charles M. Grisham

Myosin | See G&G, 2/e p. 553-554

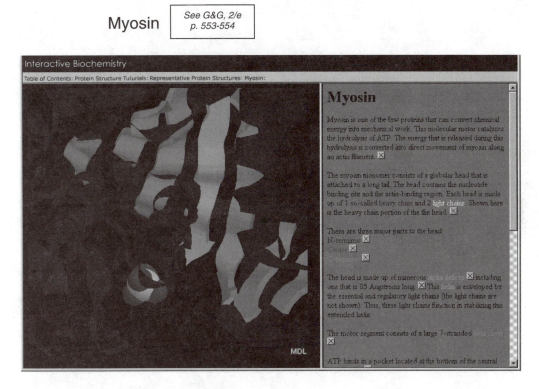

Myosin

Myosin is one of the few proteins that can convert chemical energy into mechanical work. This molecular motor catalyzes the hydrolysis of ATP. The energy that is released during this hydrolysis is converted into direct movement of myosin along an actin filament. ☒

The myosin monomer consists of a globular head that is attached to a long tail. The head contains the nucleotide binding site and the actin-binding region. Each head is made up of 1 so-called heavy chain and 2 light chains. Shown here is the heavy chain portion of the the head. ☒

There are three major parts to the head:
N-terminus ☒
Center ☒
☒

The head is made up of numerous alpha helices ☒ including one that is 85 Angstroms long. ☒ This helix is enveloped by the essential and regulatory light chains (the light chains are not shown). Thus, these light chains function in stabilizing this extended helix

The motor segment consists of a large 7-stranded ☒

ATP binds in a pocket located at the bottom of the central

MDL

1. The structure shown here represents only part of the myosin structure. Explain._____
   _____

2. Describe in words the nature and location in the head structure of the N-terminal domain, the middle or center domain, and the C-terminal domain of the heavy chain portion of the myosin head._____
   _____
   _____

3. Of all the many helices in the heavy chain portion of the myosin head, one remarkable one stands out. Why?_____

4. What is the relationship of this unique helix to the light chain proteins of myosin?____
   _____
   _____

5. A single large beta sheet is part of this structure. Where is it? What would you imagine to be its function?_____
   _____

Oxytocin

See G&G, 2/e
p. 561

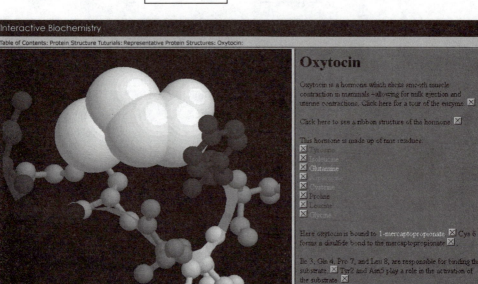

1. What is the function of oxytocin?_____

   _____

2. This is a modified oxytocin. The N-terminal amino group of the peptide is missing.
   At this position, as a result, what was a cysteine is now a β-mercaptopropionate.
   Does this modified residue participate in a disulfide bond?_____

   _____

3. What would the net charge be on oxytocin at pH 7? (assuming the presence of the N-
   terminal amino group)?_____

Charles M. Grisham

Prions

See G&G, 2/e
p. 979

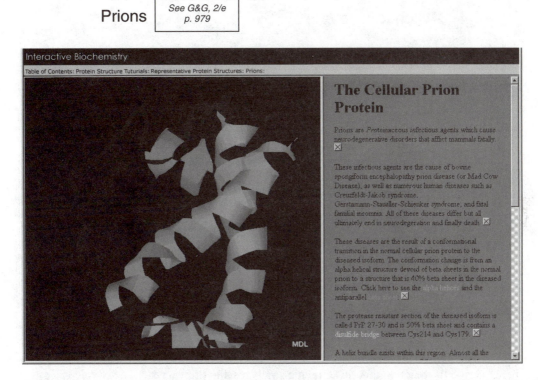

The Cellular Prion Protein

Prions are *Proteinaceous infectious agents* which cause neuro-degenerative disorders that afflict mammals fatally. ⊠

These infectious agents are the cause of bovine spongiform encephalopathy prion disease (or Mad Cow Disease), as well as numerous human diseases such as Creutzfeldt-Jakob syndrome, Gerstmann-Stausler-Schienker syndrome, and fatal familial insomnia. All of these diseases differ but all ultimately end in neurodegeration and finally death. ⊠

These diseases are the result of a conformational transition in the normal cellular prion protein to the diseased isoform. The conformation change is from an alpha helical structure devoid of beta sheets in the normal prion to a structure that is 40% beta sheet in the diseased isoform. Click here to see the alpha helices and the antiparallel ⊠

The protease resistant section of the diseased isoform is called PrP 27-30 and is 50% beta sheet and contains a disulfide bridge between Cys214 and Cys179. ⊠

A helix bundle exists within this region. Almost all the

MDL

1. How are diseases thought to be caused by prions apparently "transmitted"?_____
   _____

2. Explain the conformation difference between normal prion protein and the diseased
   isoform._____
   _____

3. Where are most prion point mutations located in the three-dimensional structure of
   the protein?_____
   _____

4. What other diseases show similarities to prion-related diseases such as mad-cow
   disease?_____
   _____

Ribonuclease A

See G&G, 2/e
p. 179

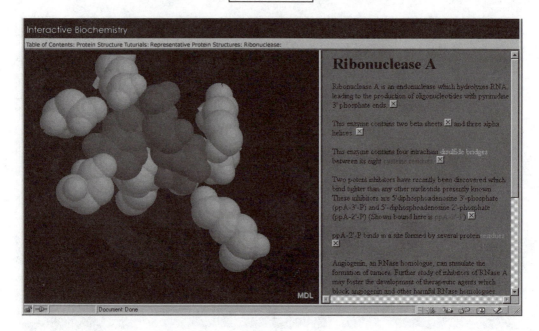

Interactive Biochemistry

Table of Contents: Protein Structure Tutorials: Representative Protein Structures: Ribonuclease:

Ribonuclease A

Ribonuclease A is an endonuclease which hydrolyzes RNA, leading to the production of oligonucleotides with pyrimidine 3' phosphate ends. ☒

This enzyme contains two beta sheets ☒ and three alpha helices ☒

This enzyme contains four intrachain disulfide bridges between its eight cysteine residues. ☒

Two potent inhibitors have recently been discovered which bind tighter than any other nucleotide presently known. These inhibitors are 5'diphosphoadenosine 3'-phosphate (ppA-3'-P) and 5'-diphosphoadenosine 2'-phosphate (ppA-2'-P) (Shown bound here is ppA-2'-P) ☒

ppA-2'-P binds in a site formed by several protein residues ☒

Angiogenin, an RNase homologue, can stimulate the formation of tumors. Further study of inhibitors of RNase A may foster the development of therapeutic agents which block angiogenin and other harmful RNase homologues.

MDL

Document Done

1. Describe the reaction that this enzyme catalyzes and draw a reaction scheme to illustrate._____

   _____

2. Study the protein carefully, and draw an approximate map of the sequence of the protein showing the locations of all four disulfide linkages.

3. Identify all the beta turns (tight turns) in this protein structure._____

4. Draw the structure of ppA-2'-P and suggest a reason why it binds so tightly to ribonuclease A.

5. Question for further reading: Look up the affinity of ppA-2'-P and compare it to the affinities of typical transition-state analogs.

Charles M. Grisham

Serpins

See G&G, 2/e
p. 194

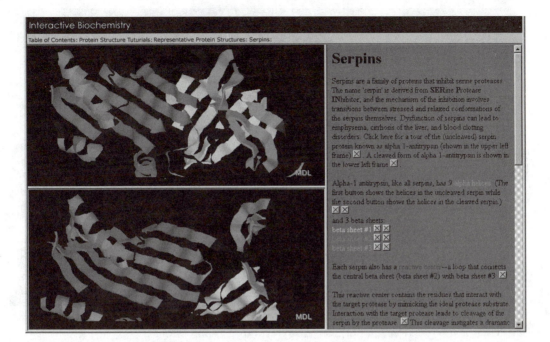

Interactive Biochemistry

Table of Contents: Protein Structure Tutorials: Representative Protein Structures: Serpins:

Serpins

Serpins are a family of proteins that inhibit serine proteases. The name 'serpin' is derived from SERine Protease INhibitor, and the mechanism of the inhibition involves transitions between stressed and relaxed conformations of the serpins themselves. Dysfunction of serpins can lead to emphysema, cirrhosis of the liver, and blood clotting disorders. Click here for a tour of the (uncleaved) serpin protein known as alpha 1-antitrypsin (shown in the upper left frame) ☒. A cleaved form of alpha 1-antitrypsin is shown in the lower left frame ☒.

Alpha-1 antitrypsin, like all serpins, has 9 alpha helices. (The first button shows the helices in the uncleaved serpin while the second button shows the helices in the cleaved serpin.) ☒ ☒
and 3 beta sheets:
beta sheet #1 ☒ ☒
beta sheet #2 ☒ ☒
beta sheet #3 ☒ ☒

Each serpin also has a reactive center—a loop that connects the central beta sheet (beta sheet #2) with beta sheet #3 ☒

This reactive center contains the residues that interact with the target protease by mimicking the ideal protease substrate. Interaction with the target protease leads to cleavage of the serpin by the protease ☒ This cleavage instigates a dramatic

1. Serpins function to keep various proteases in check. Name at least two diseases that can arise from dysfunction in a serpin or serpins._____
   _____

2. How many alpha helices and beta sheets comprise the alpha-1 antitrypsin proteins shown in this exercise?_____

3. Are the beta sheets you found parallel or antiparallel? Explain what significance this has in terms of where these helices are located in the protein structure._____
   _____
   _____

4. Define the "reactive center", both in terms of structure and in terms of function._____
   _____
   _____

5. Explain the major conformation change that occurs upon cleavage in the reactive center loop. Where does the loop go when cleavage has occurred?_____
   _____

6. How does this dramatic conformational change occur?_____
   _____

Soybean Trypsin Inhibitor

See G&G, 2/e
p. 189

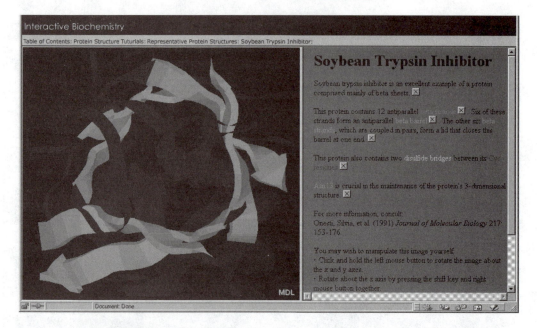

1. What is the function and purpose of soybean trypsin inhibitor?_____
_____

2. Describe the arrangement of beta strands in this protein._____
_____
_____

3. Describe the locations of all the beta turns (tight turns) that you can find in this
structure._____
_____
_____

4. Question for further reading: Consult the reference provided, and discuss the role of
Asn13 in the maintenance of structure in soybean trypsin inhibitor._____
_____
_____
_____

_____ Charles M. Grisham

Coiled Coils

GCN4

See G&G, 2/e
p. 188

GCN4

Shown here is the leucine zipper domain of the yeast transcriptional activator GCN4. ☒

Leucine zippers are peptide segments consisting of consecutive seven-residue repeat segments, in which the first and fourth residues are hydrophobic, with the fourth residue usually being leucine. Pairs of these peptides fold as short coiled coils. GCN4, a typical leucine zipper protein, forms a 2-stranded, parallel coiled coil in which the alpha helices wrap around each other in a left handed super coil. ☒

The dimer of this protein is a twisted elliptical cylinder. ☒

At the dimer interface, leucine residues make side to side interactions. ☒

Each leucine is surrounded by 4 of the neighboring helix residues. These residues greatly contribute to the stability of the dimer. ☒

The leucine zipper can be viewed as a twisted ladder. The sides of the ladder symbolize the helix backbone ☒ and the rungs of the ladder represent the leucine side chains. ☒

For more information, consult.

MDL

1. Describe the structural motif called a 7-residue repeat, which is the basis for coiled coils and leucine zipper structures._____

_____

_____

2. How are leucine residues on opposite peptides arranged with respect to each other in this structure?_____

_____

3. The twisted nature of the coiled coil depends intimately upon the nature of the alpha helix — and the fact that there are 3.6 residues per turn in the helix. What would you expect to observe in pairs of 7-residue-repeat peptides if there were precisely 3.5 residues per turn in an alpha helix?_____

_____

_____

4. Can you name other common coiled-coil proteins? Where in or on your body are coiled-coil proteins?_____

_____

_____

Hemagglutinin See G&G, 2/e p. 188

Hemagglutinin

Hemagglutinin (HA) is the surface glycoprotein of the influenza virus. This protein allows the virus to bind to cell surface conjugates by recognizing their terminal sialic acid residues. It also mediates the fusion of the virus with the endosomal membrane. ☒

HA is the variable part of the virus that is responsible for antigenic drift (resulting in recurrent influenza epidemics). ☒

HA is a trimer of identical subunits. Shown here is a hemagglutinin monomer. Each monomer is made up of 2 domains: HA₁ and ☒

HA₁ is the globular domain that contains numerous residues which form the sialic acid receptor binding site. ☒

HA₂ is the central stem structure that is helix rich. ☒

The HA trimer is a 3-stranded coiled coil, with the long helix from each of the three subunits contributing to the coiled coil structure. After endocytosis of the bound virus, the low endosomal pH dissociates the hemagglutinin stalk from the head subunits. This allows the head to refold. The refolded coiled coil propels the

MDL

1. How does the hemagglutinin protein function on the surface of influenza virus?_____
   _____
   _____

2. There are two principal domains in hemagglutinin, HA₁ and HA₂. Though they are
   not highlighted here, how many beta sheets can you find in each domain? Determine
   the number of strands in each sheet, and whether the sheets are parallel, antiparallel,
   or mixed._____
   _____
   _____
   _____

3. The most dramatic structural feature of this protein is the long alpha helix that is part
   of domain HA₂. Estimate the length of this helix by carefully counting the number of
   turns in the helix._____
   _____

Membranes and Transport

Aerolysin

See G&G, 2/e
p. 318

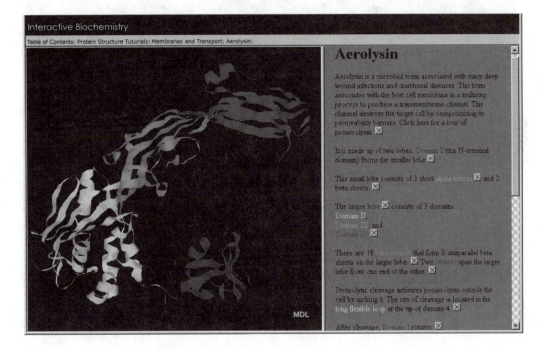

Aerolysin

Aerolysin is a microbial toxin associated with many deep wound infections and diarrhoeal diseases. This toxin associates with the host cell membrane in a multistep process to produce a transmembrane channel. This channel destroys the target cell by compromising its permeability barriers. Click here for a tour of proaerolysin ☒

It is made up of two lobes. Domain I (the N-terminal domain) forms the smaller lobe ☒

This small lobe consists of 3 short alpha helices ☒ and 2 beta sheets. ☒

The larger lobe ☒ consists of 3 domains. Domain II, Domain III, and Domain IV ☒

There are 18 ___ that form 8 antiparallel beta sheets on the larger lobe ☒ Two strands span the larger lobe from one end of the other. ☒

Proteolytic cleavage activates proaerolysin outside the cell by nicking it. The site of cleavage is located in the long flexible loop at the tip of domain 4. ☒

After cleavage, Domain I rotates ☒

MDL

1. This protein has four "domains". Which one constitutes the "small lobe" of the protein?_____

2. Which domains comprise the "large lobe"?_____

3. Which of the eight beta sheets in the large lobe are parallel and which are antiparallel?_____
   _____

4. What structural feature relating to beta sheets in this protein is extremely unusual?___
   _____

5. Where is the site of proteolytic cleavage that activates proaerolysin?_____
   _____

6. What is the function of aerolysin, and how does oligomerization affect this function?_
   _____

7. What role does domain 4 play in the function of aerolysin?_____
   _____

Bacteriorhodopsin

See G&G, 2/e
p. 273, 310

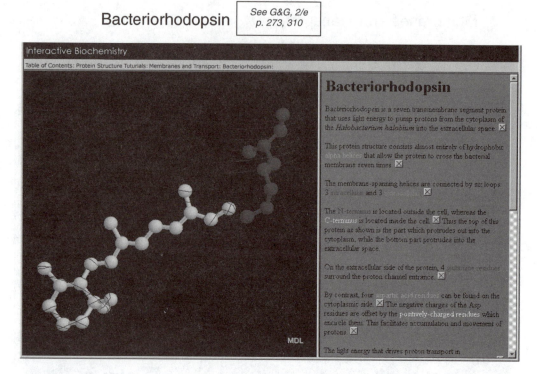

Bacteriorhodopsin

Bacteriorhodopsin is a seven transmembrane segment protein that uses light energy to pump protons from the cytoplasm of the *Halobacterium halobium* into the extracellular space. ☒

This protein structure consists almost entirely of hydrophobic alpha helices that allow the protein to cross the bacterial membrane seven times. ☒

The membrane-spanning helices are connected by six loops: 3 intracellular and 3 extracellular. ☒

The N-terminus is located outside the cell, whereas the C-terminus is located inside the cell. ☒ Thus the top of this protein as shown is the part which protrudes out into the cytoplasm, while the bottom part protrudes into the extracellular space.

On the extracellular side of the protein, 4 glutamine residues surround the proton channel entrance. ☒

By contrast, four aspartic acid residues can be found on the cytoplasmic side. ☒ The negative charges of the Asp residues are offset by the positively-charged residues which encircle them. This facilitates accumulation and movement of protons. ☒

The light energy that drives proton transport in

MDL

1. How many transmembrane segments does this protein have? How many alpha helices?_____

2. Where is the N-terminus of the protein located? Where is the C-terminus located?___
   _____

3. What cluster of residues lies at the entrance to the proton channel in bR?_____
   _____

4. What cluster of residues lies at the exit of the proton channel on the cytoplasmic face of bR?_____

5. Where is the retinal chromophore located in the bR structure?_____
   _____

6. What four amino acid residues are presumed to be involved, together with retinal, in the movement of protons through the channel?_____
   _____

7. The pK$_a$ values of Asp85 and Asp96 are remarkably high — in the vicinity of 11! Why do you suppose this is necessary, and what could possibly make the pK$_a$ values of Asp residues so high?_____
   _____

_____ Charles M. Grisham

Calcium Myristoyl Switches

See G&G, 2/e
p. 275-276

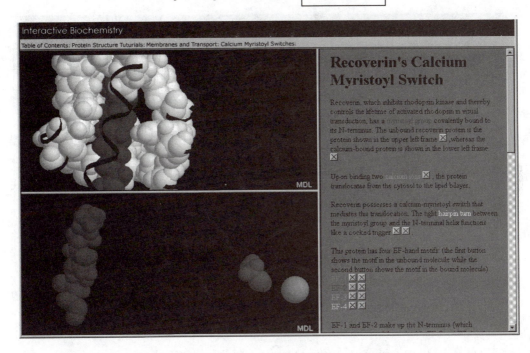

Interactive Biochemistry

Table of Contents: Protein Structure Tutorials: Membranes and Transport: Calcium Myristoyl Switches.

Recoverin's Calcium Myristoyl Switch

Recoverin, which inhibits rhodopsin kinase and thereby controls the lifetime of activated rhodopsin in visual transduction, has a myristoyl group covalently bound to its N-terminus. The unbound recoverin protein is the protein shown in the upper left frame ☒, whereas the calcium-bound protein is shown in the lower left frame ☒

Upon binding two calcium ions ☒, the protein translocates from the cytosol to the lipid bilayer.

Recoverin possesses a calcium-myristoyl switch that mediates this translocation. The tight hairpin turn between the myristoyl group and the N-terminal helix functions like a cocked trigger ☒ ☒.

This protein has four EF-hand motifs: (the first button shows the motif in the unbound molecule while the second button shows the motif in the bound molecule)

☒ ☒
☒ ☒
EF-3 ☒ ☒
EF-4 ☒ ☒

EF-1 and EF-2 make up the N-terminus (which

1. What is the "lipid anchor" group that is covalently bound to recoverin?_____
   _____

2. Where is this lipid anchor located in the Ca-free form of the protein?_____
   _____

3. Where is this lipid anchor located in the Ca-bound form of the protein?_____
   _____

4. What part of the protein chain is instrumental in effecting this conversion?_____
   _____

5. Which of the EF-hand domains receive the Ca^{2+} ions when they bind to the protein?_
   _____

6. What motifs in the protein coordinate the myristoyl anchor in the Ca-free form of the
   protein?_____

7. Rotations about two glycine residues facilitate the massive conformation change
   here. Which glycines are they, and why do you suppose glycines are involved?_____
   _____

Colicin See G&G, 2/e p. 315-316

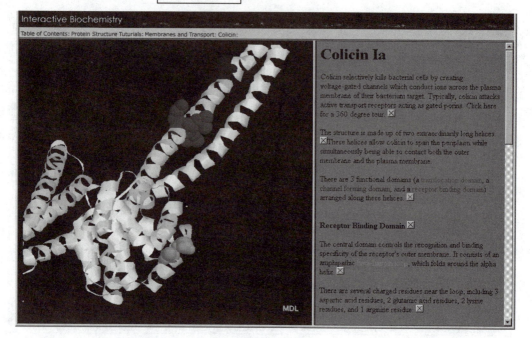

Colicin Ia

Colicin selectively kills bacterial cells by creating voltage-gated channels which conduct ions across the plasma membrane of their bacterium target. Typically, colicin attacks active transport receptors acting as gated porins. Click here for a 360 degree tour. ☒

The structure is made up of two extraordinarily long helices. ☒ These helices allow colicin to span the periplasm while simultaneously being able to contact both the outer membrane and the plasma membrane.

There are 3 functional domains (a translocation domain, a channel forming domain, and a receptor binding domain) arranged along these helices. ☒

Receptor Binding Domain ☒

The central domain controls the recognition and binding specificity of the receptor's outer membrane. It consists of an amphipathic _____, which folds around the alpha helix ☒

There are several charged residues near the loop, including 3 aspartic acid residues, 2 glutamic acid residues, 2 lysine residues, and 1 arginine residue. ☒

MDL

1. What is the most striking structural feature of this protein? What functional requirement is facilitated by the unusual feature?_____
_____

2. What structural feature, wrapped around an alpha helix in the receptor-binding domain, is involved in controlling recognition and binding specificity at the outer membrane?_____

3. What is the colicin TonB box, and what does it do?_____
_____

4. What is the "hydrophobic hairpin loop" in the channel-forming domain, and how does it function in channel formation?_____
_____

5. Describe the nature and chemical character of the eight helices surrounding the hydrophobic hairpin loop._____
_____

6. Though most of the protein is normally protected from proteolysis (by lipid) in the natural state, several residues are in fact subject to proteolysis. What are they?_____
_____
_____

_____ Charles M. Grisham

Delta Endotoxin

See G&G, 2/e
p. 317

Delta Endotoxin

Delta Endotoxins are toxins which are specifically lethal to *Lepidoptera*, *Diptera*, and *Cleopatera* insects and have been formulated into commercial insecticides. They are synthesized as crystallized proteins. Once ingested by the insect, the crystal is dissolved by the alkaline pH in the mid-gut and cleaved by proteases found in the gut. Now an active toxin, the endotoxin binds to protein receptors in the epithelium of the mid-gut. This results in membrane lesions that lead to swelling and lysis of the gut epithelium. Eventually, death results from starvation and septicaemia. Click here for a tour of the active endotoxin. ☒

This toxin has three domains:
☒ An N-terminal domain,
☒ A β sheet domain, and
☒ A beta sandwich domain.

The N-terminal domain consists of a seven-helix bundle. ☒ A central hydrophobic helix ☒ is surrounded by six amphipathic helices (which are tilted 20 degrees with respect to the hydrophobic helix). ☒ These helices are long enough to span the thick hydrophobic region of a membrane bilayer.

When the toxin binds to the epithelium, conformational changes occur that allow the insertion of a hairpin into the membrane. This insertion creates a defect in the membrane and allows the rest of the N-terminal domain to contribute to

1. What are the three domains found in this protein?_____
   _____

2. Describe the N-terminal domain of this protein and the structural features that
   probably facilitate pore formation and cell lysis by this toxin._____
   _____

3. Which domain of the protein is responsible for receptor binding?_____

4. What is the character of the core of this domain?_____

5. Which segment of this domain determines specificity of the endotoxin?_____
   _____

6. Are the beta sheets of the beta-sandwich domain parallel or antiparallel?_____

7. From what you know about parallel and antiparallel beta sheets, what would you
   expect to find in the core of the beta-sandwich domain, and what would you expect
   to find on the surface?_____
   _____
   _____

Hemolysin

See G&G, 2/e
p. 317

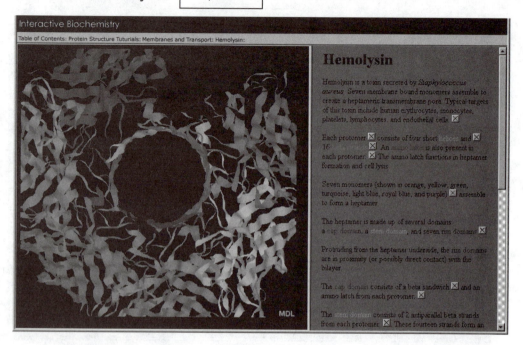

Hemolysin

Hemolysin is a toxin secreted by *Staphylococcus aureus*. Seven membrane bound monomers assemble to create a heptameric transmembrane pore. Typical targets of this toxin include human erythrocytes, monocytes, platelets, lymphocytes, and endothelial cells.

Each protomer consists of four short helices and 16. An amino latch is also present in each protomer. The amino latch functions in heptamer formation and cell lysis.

Seven monomers (shown in orange, yellow, green, turquoise, light blue, royal blue, and purple) assemble to form a heptamer.

The heptamer is made up of several domains: a cap domain, a stem domain, and seven rim domains.

Protruding from the heptamer underside, the rim domains are in proximity (or possibly direct contact) with the bilayer.

The cap domain consists of a beta sandwich and an amino latch from each protomer.

The stem domain consists of 2 antiparallel beta strands from each protomer. These fourteen strands form an

MDL

1. Hemolysin from *Staphylococcus aureus* is composed of seven identical subunits or "protomers". What elements of secondary structure can you find in the protomer structure?_____

_____

2. When seven protomers assemble to form a heptamer channel, what three domains can be defined?_____

_____

3. Which of these three are (or could be) in contact with the membrane bilayer when channel insertion has occurred?_____

4. What element of secondary structure is the core of the cap domain?_____

_____

5. The stem domain is a heptameric assembly of what kind of structural elements provided by each of the protomers?_____

6. What kind of beta barrel is formed in the stem domain?_____

7. Where are hydrophobic residues located in the stem domain?_____

8. What three amino acids from each protomer form a constriction in the channel?_____

_____

_____ Charles M. Grisham

Maltoporin

See G&G, 2/e
p. 274, 313-315

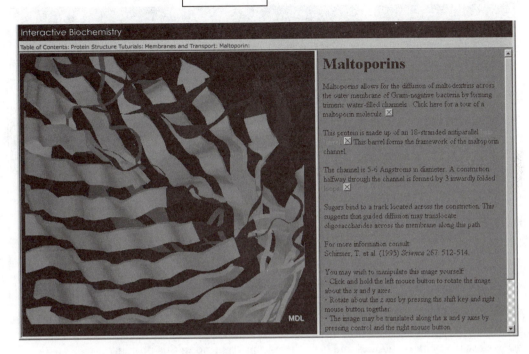

Interactive Biochemistry

Table of Contents: Protein Structure Tutorials: Membranes and Transport: Maltoporin:

Maltoporins

Maltoporins allows for the diffusion of malto-dextrins across the outer membrane of Gram-negative bacteria by forming trimeric water-filled channels. Click here for a tour of a maltoporin molecule ☒

This protein is made up of an 18-stranded antiparallel ☒ This barrel forms the framework of the maltoporin channel.

The channel is 5-6 Angstroms in diameter. A constriction halfway through the channel is formed by 3 inwardly folded loops ☒

Sugars bind to a track located across the constriction. This suggests that guided diffusion may translocate oligosaccharides across the membrane along this path.

For more information consult:
Schirmer, T. et al. (1995) *Science* 267: 512-514.

You may wish to manipulate this image yourself:
· Click and hold the left mouse button to rotate the image about the x and y axes.
· Rotate about the z axis by pressing the shift key and right mouse button together.
· The image may be translated along the x and y axes by pressing control and the right mouse button.

MDL

1. What is the normal function of maltoporin?_____
_____

2. What is the nature of the barrel structure formed by maltoporin?_____
_____

3. What is the diameter of the channel formed by the barrel?_____
_____

4. What forms the constriction in the channel? _____
_____

5. What is the apparent role of this constriction?_____
_____

Ras: a GTPase Enzyme

See G&G, 2/e p. S-9

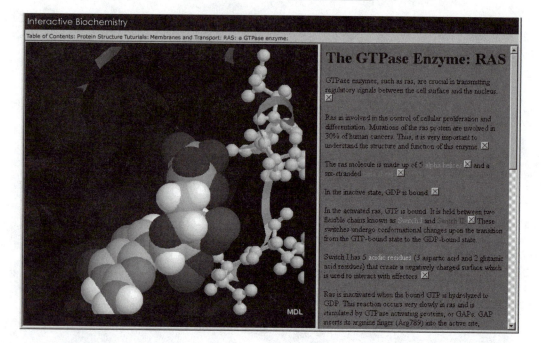

Interactive Biochemistry

Table of Contents: Protein Structure Tutorials: Membranes and Transport: RAS: a GTPase enzyme:

The GTPase Enzyme: RAS

GTPase enzymes, such as ras, are crucial in transmitting regulatory signals between the cell surface and the nucleus. ☒

Ras in involved in the control of cellular proliferation and differentiation. Mutations of the ras protein are involved in 30% of human cancers. Thus, it is very important to understand the structure and function of this enzyme. ☒

The ras molecule is made up of 5 alpha helices ☒ and a six-stranded ☒

In the inactive state, GDP is bound. ☒

In the activated ras, GTP is bound. It is held between two flexible chains known as Switch I and Switch II. ☒ These switches undergo conformational changes upon the transition from the GTP-bound state to the GDP-bound state.

Switch I has 5 acidic residues (3 aspartic acid and 2 glutamic acid residues) that create a negatively charged surface which is used to interact with effectors ☒

Ras is inactivated when the bound GTP is hydrolyzed to GDP. This reaction occurs very slowly in ras and is stimulated by GTPase activating proteins, or GAPs. GAP inserts its arginine finger (Arg789) into the active site,

MDL

1. How many alpha helices and beta sheets can you find in this protein?_____
_____

2. What kind(s) of beta sheet(s)?_____

3. What is the nucleotide that is found in this structure?_____

4. What are Switch I and Switch II, and what is their function?_____
_____
_____

5. What is the origin of the negatively charged surface on Switch I?_____
_____

6. What is the role of Gln61 in Switch II?_____
_____

7. Why are essentially all mutations at Gly12 oncogenic?_____
_____
_____

_____ Charles M. Grisham

Molecular Motors

ATP Synthase

See G&G, 2/e p. 695

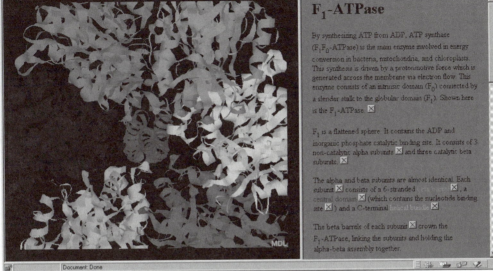

F₁-ATPase

By synthesizing ATP from ADP, ATP synthase (F_1F_0-ATPase) is the main enzyme involved in energy conversion in bacteria, mitochondria, and chloroplasts. This synthesis is driven by a protonmotive force which is generated across the membrane via electron flow. This enzyme consists of an intrinsic domain (F_0) connected by a slender stalk to the globular domain (F_1). Shown here is the F_1-ATPase. ☒

F_1 is a flattened sphere. It contains the ADP and inorganic phosphate catalytic binding site. It consists of 3 non-catalytic alpha subunits ☒ and three catalytic beta subunits. ☒

The alpha and beta subunits are almost identical. Each subunit ☒ consists of a 6-stranded ☒, a central domain ☒ (which contains the nucleotide binding site ☒) and a C-terminal helical bundle. ☒

The beta barrels of each subunit ☒ crown the F_1-ATPase, linking the subunits and holding the alpha-beta assembly together.

Document: Done

---

1. What three distinct domains can be found in each of the alpha and beta subunits?_____
   _____
   _____

2. Which of these contains the nucleotide-binding site?_____
   _____

3. Where in the assembled F_1 structure are the beta barrels located?_____
   _____

4. The nucleotide-binding domain is a classic structure with a nine-stranded beta sheet and several alpha helices. Is the sheet parallel or antiparallel? Where are the helices?
   _____
   _____

5. What residues define the nucleotide-binding pocket?_____
   _____

6. What is the molecular explanation for the fact that the alpha subunits, though highly similar to the betas, are noncatalytic?_____
   _____

7. Describe the interactions between the gamma subunit and the alpha and beta subunits.

_____

Myosin (See page 51)

Enzyme Mechanisms

Alcohol Dehydrogenase See G&G, 2/e p. 656-657

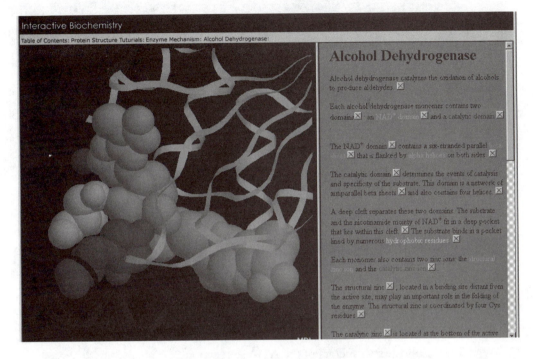

1. The alcohol dehydrogenase (ADH) shown here is that from equine liver. Describe the structure and function of the two domains of equine liver ADH._____

   _____

   _____

2. Where do the substrate and the NAD cofactor bind to the enzyme?_____

   _____

3. Where do the two zinc ions bind, and what are their respective roles?_____

   _____

4. The alcohol dehydrogenase reaction involves proton abstraction from the -OH group and hydride removal from the adjacent carbon. Which residues assist in proton abstraction?_____

5. What is the NAD(H) analog that is bound in this structure, what is its effect on the enzyme, and how does it exert this effect, in terms of structural alterations at the active site?_____

   _____

See G&G, 2/e
p. 914-916

ATCase

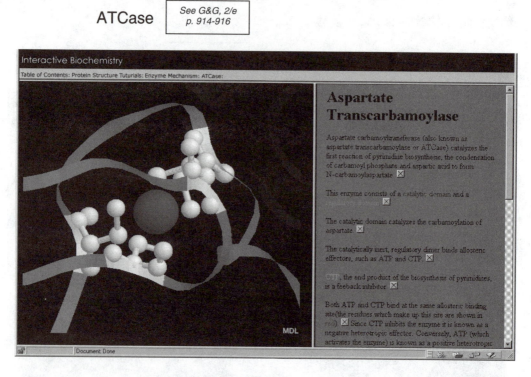

Aspartate Transcarbamoylase

Aspartate carbamoyltransferase (also known as aspartate transcarbamoylase or ATCase) catalyzes the first reaction of pyrimidine biosynthesis, the condensation of carbamoyl phosphate and aspartic acid to form N-carbamoylaspartate ☒

This enzyme consists of a catalytic domain and a _____ ☒

The catalytic domain catalyzes the carbamoylation of aspartate ☒

The catalytically inert, regulatory dimer binds allosteric effectors, such as ATP and CTP. ☒

CTP, the end product of the biosynthesis of pyrimidines, is a feedback inhibitor ☒

Both ATP and CTP bind at the same allosteric binding site(the residues which make up this site are shown in red) ☒ Since CTP inhibits the enzyme it is known as a negative heterotropic effector. Conversely, ATP (which activates the enzyme) is known as a positive heterotropic

MDL

Document: Done

1. What is the metabolic function of ATCase?_____

_____

2. What are the allosteric effectors of the ATCase reaction?_____

_____

3. What is the feedback inhibitor of ATCase?_____

4. Define the terms "positive heterotropic effector" and "negative heterotropic effector".

_____

_____

5. Which residues move upon ATP binding, so that the nucleotide binding site is enlarged?_____

6. What metal ion binds to the regulatory domain, and what amino acid residues coordinate it?_____

7. Which amino acid residue is thought to mediate communication between the metal site and the allosteric regulatory site?_____

Chymotrypsin (See page 37)

Glycogen Phosphorylase
See G&G, 2/e
p. 473-479

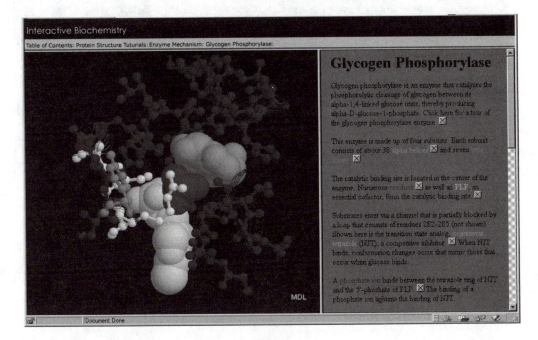

1. The glycogen phosphorylase (GP) reaction is a phosphorolysis. What does this mean?_____

2. How many alpha helices and beta sheets can you find in a glycogen phosphorylase monomer?_____

3. What coenzyme binds to GP to form part of the active site?_____

4. Which residues of the protein constitute the loop that blocks substrate binding?_____
   _____

5. What is the transition state analog whose binding induces conformation changes that mimic those that occur upon glucose binding?_____

6. What effect does phosphate binding have on the affinity for nojirimycin tetrazole?____
   _____

7. Which amino acid residues create a suitable binding site for the 5'-phosphate of PLP?
   _____

8. Where is the caffeine binding site located on the enzyme?_____

Haloacid Dehalogenase

See G&G, 2/e
p. 301-307

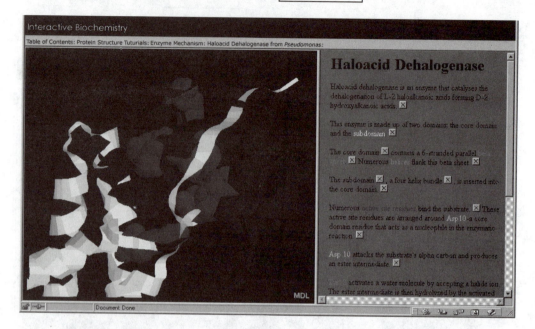

1. Consult the reference provided in the tutorial, and write an equation for the reaction catalyzed by this enzyme.

2. This structure of the core domain is a classic parallel beta sheet with helices on either side of the sheet. Think about the nature of parallel beta sheets and the requirement that the surface of this protein must interact with solvent on its surface, and describe the character of these alpha helices._____

_____

_____

3. How is the subdomain inserted into the core domain?_____

_____

4. Consult the reference provided and write a suitable mechanism for this reaction, invoking Asp10 and Tyr12, as described in the tutorial exercise.

    HIV-1 Protease (See page 46)
    Lysozyme (See page 48)
    Serpins (See page 55)

Charles M. Grisham

Enzymes of Metabolism

Aconitase

See G&G, 2/e
p. 648-650

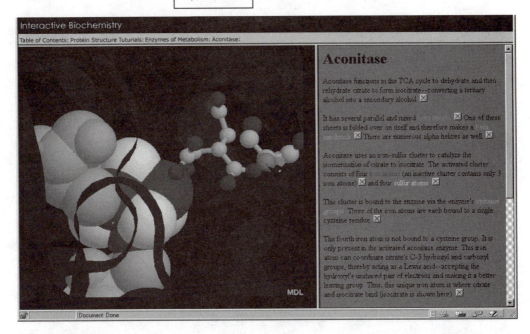

1. The reaction catalyzed by aconitase is a chemically challenging one. Describe it, both in terms of the general chemistry of the reaction and in terms of the specific substrate and product of the reaction._____

_____

2. How many beta sheets can you find in this structure? Are they parallel, antiparallel, or mixed?_____

_____

3. How many Fe atoms are bound in this particular structure? How many are required for activation of aconitase?_____

4. Describe the behavior of the fourth Fe atom in the activity of this enzyme._____

_____

_____

Apolipoprotein A-1 (See page 36)

Citrate Synthase

See G&G, 2/e
p. 644-645

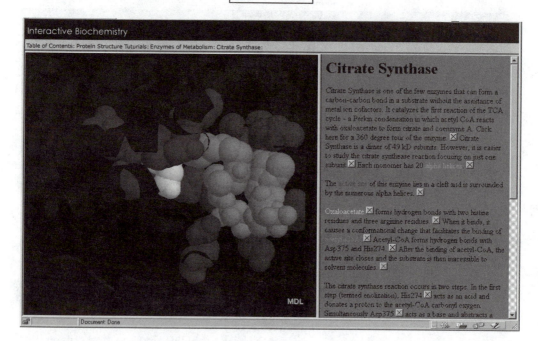

Interactive Biochemistry

Table of Contents: Protein Structure Tutorials: Enzymes of Metabolism: Citrate Synthase:

Citrate Synthase

Citrate Synthase is one of the few enzymes that can form a carbon-carbon bond in a substrate without the assistance of metal ion cofactors. It catalyzes the first reaction of the TCA cycle - a Perkin condensation in which acetyl CoA reacts with oxaloacetate to form citrate and coenzyme A. Click here for a 360 degree tour of the enzyme ⊠ Citrate Synthase is a dimer of 49 kD subunits. However, it is easier to study the citrate synthase reaction focusing on just one subunit ⊠ Each monomer has 20 alpha helices. ⊠

The active site of this enzyme lies in a cleft and is surrounded by the numerous alpha helices. ⊠

Oxaloacetate ⊠ forms hydrogen bonds with two histine residues and three arginine residues. ⊠ When it binds, it causes a conformational change that facilitates the binding of ⊠ Acetyl-CoA forms hydrogen bonds with Asp375 and His274 ⊠ After the binding of acetyl-CoA, the active site closes and the substrate is then inaccessible to solvent molecules ⊠

The citrate synthase reaction occurs in two steps. In the first step (termed enolization), His274 ⊠ acts as an acid and donates a proton to the acetyl-CoA carbonyl oxygen. Simultaneously Asp375 ⊠ acts as a base and abstracts a

MDL

Document Done

1. The first reaction of the TCA cycle, catalyzed by citrate synthase (CS), is a Perkin condensation. What is a Perkin condensation?_____
   _____

2. How many alpha helices can you find in a monomer of citrate synthase?_____

3. What effect does oxaloacetate binding have on acetyl-CoA binding?_____
   _____

4. In the CS reaction, what residue acts as a general acid, donating a proton to the carbonyl oxygen of acetyl-CoA?_____

5. What residue acts as a general base, abstracting a proton from the methyl group of acetyl-CoA?_____

6. Which residue protonates the carbonyl oxygen of oxaloacetate in the second phase of the reaction?_____

Charles M. Grisham

Cyclooxygenase

See G&G, 2/e
p. 832-834

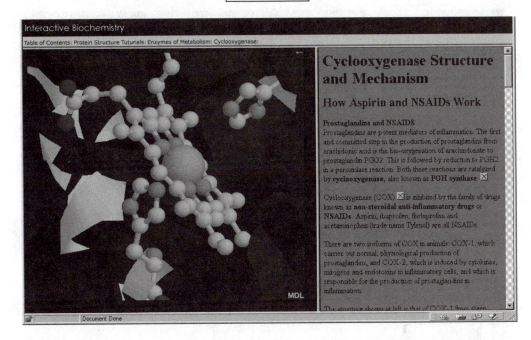

1. Describe the properties of the two isoforms of cyclooxygenase, COX-1 and COX-2.__
    _____
    _____

2. Which residues comprise the membrane-binding motif, and what is its structure?_____
    _____

3. Which residues provide the axial ligands of the Fe in the heme group?_____
    _____

4. Which residues comprise the alpha helices that constitute the walls of the substrate
    tunnel?_____

5. Which residue forms a radical that is believed essential for the cyclooxygenase
    activity?_____

6. From what you have learned in this exercise, summarize the molecular events of the
    therapeutic action of NSAIDs._____
    _____

7. Several specific COX-2 inhibitors are now on the market or are entering clinical
    trials. Why is this important news?_____
    _____

Enolase

See G&G, 2/e p. 628-629

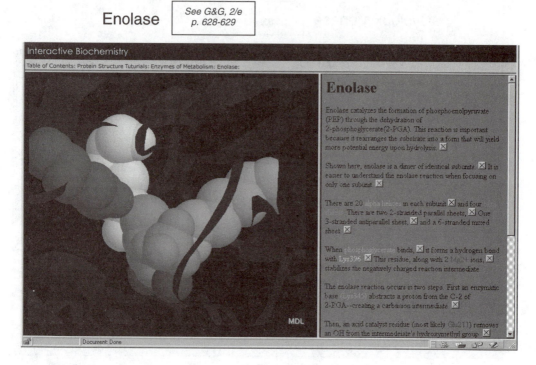

Table of Contents: Protein Structure Tuturials: Enzymes of Metabolism: Enolase:

Enolase

Enolase catalyzes the formation of phosphoenolpyruvate (PEP) through the dehydration of 2-phosphoglycerate (2-PGA). This reaction is important because it rearranges the substrate into a form that will yield more potential energy upon hydrolysis.

Shown here, enolase is a dimer of identical subunits. It is easier to understand the enolase reaction when focusing on only one subunit.

There are 20 alpha helices in each subunit and four There are two 2-stranded parallel sheets. One 3-stranded antiparallel sheet, and a 6-stranded mixed sheet.

When phosphoglycerate binds, it forms a hydrogen bond with Lys396. This residue, along with 2 Mg2+ ions, stabilizes the negatively charged reaction intermediate

The enolase reaction occurs in two steps. First an enzymatic base (Lys345) abstracts a proton from the C-2 of 2-PGA--creating a carbanion intermediate

Then, an acid catalyst residue (most likely Glu211) removes an OH from the intermediate's hydroxymethyl group

MDL

Document: Done

1. What reaction does enolase catalyze?_____
   _____

2. How many alpha helices can you identify in an enolase monomer?_____

3. Describe each of the beta sheets found in this enzyme?_____
   _____

4. What amino acid side chain forms a hydrogen bond with the 2-phosphoglycerate substrate?_____

5. Which residue on the enzyme acts as a general base to abstract a proton from the substrate, forming the carbanion required for this reaction?_____

6. Draw the reaction in which the residue you named in question 5 abstracts the proton to form the required carbanion.

7. Draw the next step in the reaction — the production of PEP — in which an acidic residue donates a proton to facilitate formation of water from the hydroxymethyl group at C-3.

_____ Charles M. Grisham

Hexokinase

See G&G, 2/e
p. 613-616

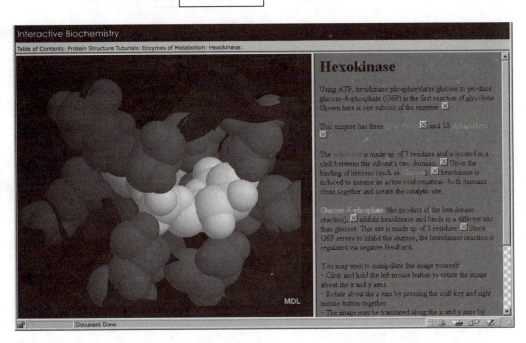

Hexokinase

Using ATP, hexokinase phosphorylates glucose to produce glucose-6-phosphate (G6P) in the first reaction of glycolysis. Shown here is one subunit of the enzyme.

This enzyme has three ___ and 13 alpha-helices.

The active site is made up of 7 residues and is located in a cleft between this subunit's two domains. Upon the binding of hexoses (such as Glucose), hexokinase is induced to assume an active conformation--both domains close together and create the catalytic site.

Glucose-6-phosphate (the product of the hexokinase reaction), inhibits hexokinase and binds in a different site than glucose. This site is made up of 8 residues. Since G6P serves to inhibit the enzyme, the hexokinase reaction is regulated via negative feedback.

You may wish to manipulate this image yourself:
- Click and hold the left mouse button to rotate the image about the x and y axes.
- Rotate about the z axis by pressing the shift key and right mouse button together.
- The image may be translated along the x and y axes by

MDL

Document: Done

1. What is the chemical logic of making glucose-6-P in this reaction? In other words, what does this reaction gain for the strategy of glycolysis?_____
   _____

2. For each of the three beta sheets in this protein, describe the number of strands and the kind of sheet._____
   _____

3. Glucose is shown bound to the active site. What other substrate must also bind before the hexokinase reaction can occur?_____

4. What is the nature of the conformational change that occurs when both substrates are bound?_____

5. What allosteric regulator is shown bound to hexokinase here?_____

6. What is the name for inhibition of this enzyme by glucose-6-P?_____

7. What are the standard-state free-energy changes for (a) the phosphorylation of glucose, and (b) the hexokinase reaction?_____

8. Look up an example of cellular concentrations of ATP, ADP, glucose, and G-6-P, and calculate the actual cellular free-energy change that occurs thanks to this enzyme.

Isocitrate Dehydrogenase

See G&G, 2/e
p. 651

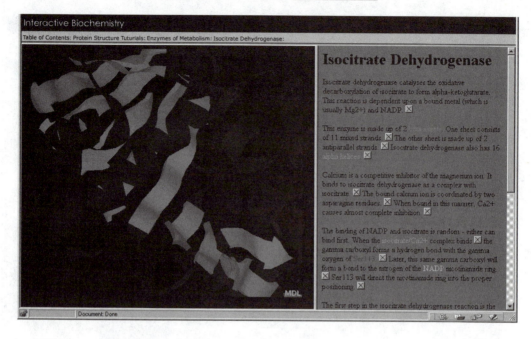

1. Write an equation which shows that this reaction is an oxidative decarboxylation._____

_____

2. Where is the mixed beta sheet in this enzyme structure?_____

_____

3. Alpha helices cover the mixed beta sheet on both sides. What kinds of amino acid side chains would you expect to find on either side of the beta sheet (i.e., between the helices and the sheet)?_____

4. Draw a diagram to illustrate the formation of a hydrogen bond between isocitrate and Ser113.

5. Which protein residues have been suggested as the general base that abstracts a proton from the isocitrate hydroxyl group to initiate this reaction?_____

_____

Charles M. Grisham

Malate Dehydrogenase

See G&G, 2/e
p. 655-658

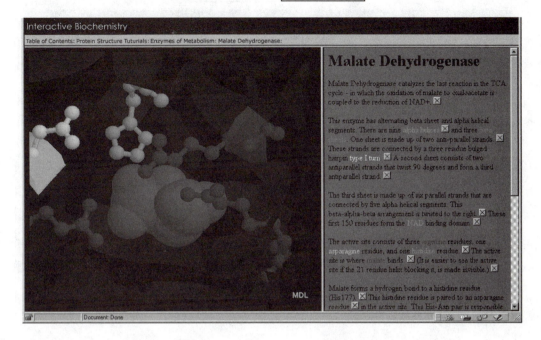

Malate Dehydrogenase

Malate Dehydrogenase catalyzes the last reaction in the TCA cycle - in which the oxidation of malate to oxaloacetate is coupled to the reduction of NAD+. ☒

This enzyme has alternating beta sheet and alpha helical segments. There are nine alpha helices ☒ and three . One sheet is made up of two anti-parallel strands ☒ These strands are connected by a three residue bulged hairpin type I turn ☒ A second sheet consists of two antiparallel strands that twist 90 degrees and form a third antiparallel strand. ☒

The third sheet is made up of six parallel strands that are connected by five alpha helical segments. This beta-alpha-beta arrangement is twisted to the right. ☒ These first 150 residues form the NAD binding domain. ☒

The active site consists of three arginine residues, one asparagine residue, and one histidine residue. ☒ The active site is where malate binds ☒ (It is easier to see the active site if the 21 residue helix blocking it, is made invisible.) ☒

Malate forms a hydrogen bond to a histidine residue (His 177) ☒ This histidine residue is paired to an asparagine residue ☒ in the active site. This His-Asn pair is responsible

MDL

Document: Done

---

1. How many alpha helices can you identify in this structure?_____

2. Describe each of the beta sheets in this protein._____
_____

3. Can you find any beta turns in this structure?_____

4. Which amino acid forms a hydrogen bond with bound malate? Which two residues work together to effect proton abstraction from the malate C-2 hydroxyl?_____
_____

5. Position C-4 of the nicotinamide ring of NAD lies in van der Waals contact with which hydrogen of malate, facilitating hydride transfer?_____

Phosphofructokinase

See G&G, 2/e p. 617-619

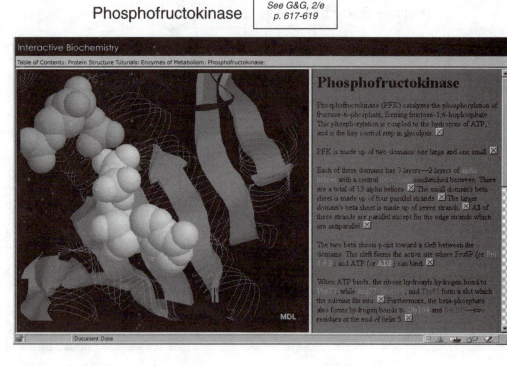

Phosphofructokinase

Phosphofructokinase (PFK) catalyzes the phosphorylation of fructose-6-phosphate, forming fructose-1,6-bisphosphate. This phosphorylation is coupled to the hydrolysis of ATP, and is the key control step in glycolysis.

PFK is made up of two domains: one large and one small.

Each of these domains has 3 layers—2 layers of alpha helices with a central sandwiched between. There are a total of 13 alpha helices. The small domain's beta sheet is made up of four parallel strands. The larger domain's beta sheet is made up of seven strands. All of these strands are parallel except for the edge strands which are antiparallel.

The two beta sheets point toward a cleft between the domains. This cleft forms the active site where Fru6P (or Fru 1,6 P) and ATP (or ADP) can bind.

When ATP binds, the ribose hydroxyls hydrogen bond to Phe, while Arg77 and Tyr41 form a slot which the adenine fits into. Furthermore, the beta-phosphate also forms hydrogen bonds to Gly104 and Ser106—two residues at the end of helix 5.

MDL

Document: Done

1. Write the reaction catalyzed by phosphofructokinase (PFK).

2. Each of the two domains of PFK consists of three "layers". Explain in detail._____
   _____
   _____

3. Are the beta sheets in each of these domains parallel or antiparallel? Why should the difference matter to the structure of the protein? Explain._____
   _____

4. Where do the substrates (fructose-6-P and ATP) bind to the enzyme?_____
   _____

5. Which amino acid residues coordinate the bound nucleotide?_____
   _____

6. Describe the coordination of fructose-6-P (or fructose-1,6-bisP) to the enzyme._____
   _____
   _____

7. Describe the binding of ADP to the allosteric effector site._____
   _____

Phosphoglycerate Kinase

See G&G, 2/e
p. 624-626

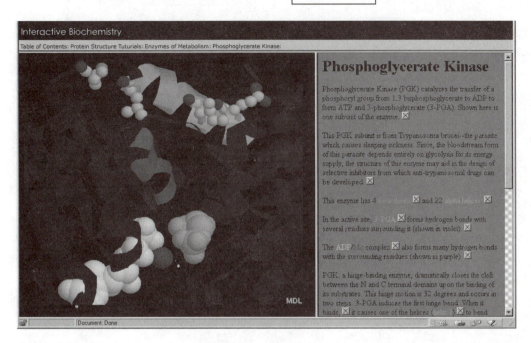

1. Discuss the nature of the beta sheets in phosphoglycerate kinase (PGK), including the layer structure of the protein and the exposure to solvent._____

   _____

2. Where are the substrate binding sites on the enzyme? Name at least one other metabolic enzyme that binds substrates in a similar location._____

   _____

3. What event initiates hinge bending to close the active site cleft?_____

   _____

4. How large (in degrees) is the first stage of hinge bending, and how large is the total hinge bending?_____

5. What initiates the second phase of hinge bending?_____

Phosphoglycerate Mutase

See G&G, 2/e
p. 626-628

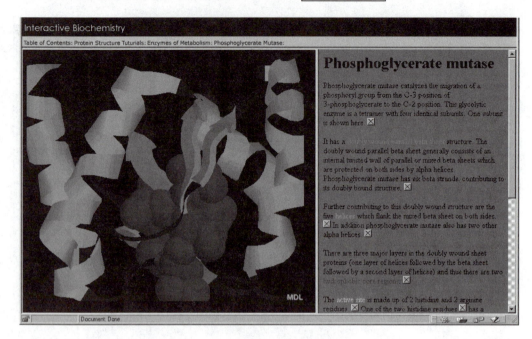

Interactive Biochemistry

Table of Contents: Protein Structure Tutorials: Enzymes of Metabolism: Phosphoglycerate Mutase:

Phosphoglycerate mutase

Phosphoglycerate mutase catalyzes the migration of a phosphoryl group from the C-3 position of 3-phosphoglycerate to the C-2 position. This glycolytic enzyme is a tetramer with four identical subunits. One subunit is shown here. ☒

It has a _____ structure. The doubly wound parallel beta sheet generally consists of an internal twisted wall of parallel or mixed beta sheets which are protected on both sides by alpha helices. Phosphoglycerate mutase has six beta strands, contributing to its doubly bound structure. ☒

Further contributing to this doubly wound structure are the five helices which flank the mixed beta sheet on both sides. ☒ In addition phosphoglycerate mutase also has two other alpha helices ☒

There are three major layers in the doubly wound sheet proteins (one layer of helices followed by the beta sheet followed by a second layer of helices) and thus there are two hydrophobic core regions ☒

The active site is made up of 2 histidine and 2 arginine residues. ☒ One of the two histidine residues ☒ has a

MDL

Document: Done

1. What is the reaction catalyzed by phosphoglycerate mutase (PGM)?_____
   _____

2. This reaction merely moves a phosphate from one carbon to another. What is the chemical purpose of this reaction?_____
   _____

3. What is the core structure of the PGM monomer?_____
   _____

4. On which side(s) of the six-stranded beta sheet would you expect to find hydrophobic residues?_____

5. What amino acid residues play a role at the active site?_____
   _____

6. How do these residues assist in the reaction that occurs at the active site?_____
   _____
   _____

Charles M. Grisham

Pyruvate Kinase

See G&G, 2/e
p. 629-630

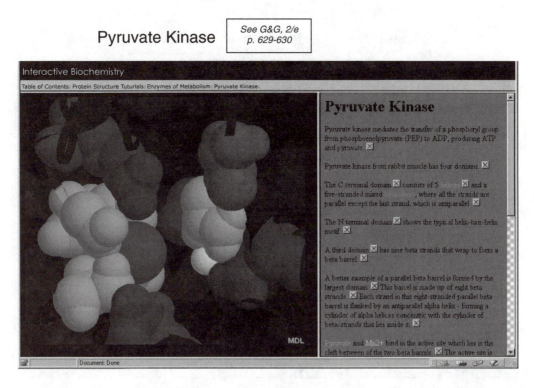

Interactive Biochemistry

Table of Contents: Protein Structure Tutorials: Enzymes of Metabolism: Pyruvate Kinase:

Pyruvate Kinase

Pyruvate kinase mediates the transfer of a phosphoryl group from phosphoenolpyruvate (PEP) to ADP, producing ATP and pyruvate. ☒

Pyruvate kinase from rabbit muscle has four domains. ☒

The C-terminal domain ☒ consists of 5 helices ☒ and a five-stranded mixed [ ], where all the strands are parallel except the last strand, which is antiparallel ☒

The N-terminal domain ☒ shows the typical helix-turn-helix motif ☒

A third domain ☒ has nine beta strands that wrap to form a beta barrel. ☒

A better example of a parallel beta barrel is formed by the largest domain. ☒ This barrel is made up of eight beta strands. ☒ Each strand in this eight-stranded parallel beta barrel is flanked by an antiparallel alpha helix - forming a cylinder of alpha helices concentric with the cylinder of beta-strands that lies inside it. ☒

Pyruvate and Mn2+ bind in the active site which lies in the cleft between of the two beta barrels. ☒ The active site is

MDL

Document: Done

1. The pyruvate kinase reaction produces a high-energy phosphoric anhydride, ATP. What makes this energy-costly reaction thermodynamically feasible?_____

_____

2. The C-terminal domain of the enzyme from rabbit muscle has a mostly parallel beta sheet with one antiparallel strand. Notice where this latter strand is located, and speculate on its composition._____

_____

_____

3. Compare and contrast the structures of the two beta barrels found in this structure.____

_____

_____

4. What could you predict about the character and nature of the eight alpha helices that surround the parallel beta barrel?_____

_____

5. Where is the active site of pyruvate kinase located?_____

_____

6. Describe the functions of the amino acids in the active site of pyruvate kinase._____

_____

_____

Triose Phosphate Isomerase

See G&G, 2/e
p. 620-621

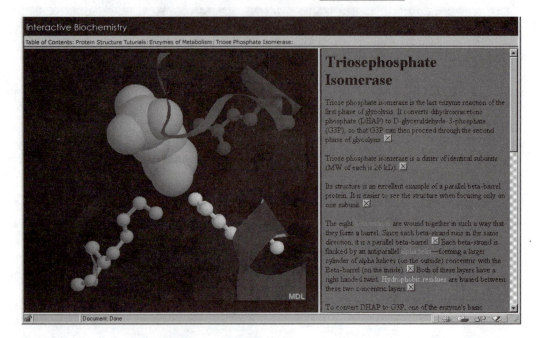

1. What is the core structure of this enzyme?_____
   _____

2. From what you know about protein structure, examine this structure and then describe
   the likely character of the alpha helices that surround the beta barrel._____
   _____

3. What roles do His95 and Lys12 play in the binding of substrate?_____
   _____

4. Which residue apparently abstracts a proton from the C-1 position of the substrate
   dihydroxyacetone phosphate?_____

5. Draw the structure of phosphoglycolate, an apparent transition-state analog for the
   TPI reaction.

6. Describe the events that occur at the active site when phosphoglycolate binds._____
   _____
   _____

Charles M. Grisham

Toxin Proteins
    Aerolysin (See page 59)

    Cholera Toxin See G&G, 2/e
p. 125, S-7, S-8

1. Describe the association of subunit A of cholera toxin (CT) with subunit B._____
   _____

2. Cholera toxin binds to oligosaccharides on cell surfaces. The binding of a
   pentasaccharide in this crystal structure imitates this interaction. Describe the binding
   of pentasaccharide to the B subunit of CT._____
   _____

3. What entity caps the beta barrel in each B subunit?_____

4. Identify the charged residues to facilitate the formation of salt bridges._____
   _____

5. Fragments A1 and A2 of the A subunit are derived from a single peptide precursor.
   What is their only linkage in the mature toxin?_____

6. What enzyme activities are exhibited by fragment A1?_____
   _____

7. Describe the A2 helix in terms of its structure and function. _____
_____
_____

_____

Colicin (See page 62)
Delta Endotoxin (See page 63)

Diphtheria Toxin | See G&G, 2/e p. 1115-1116

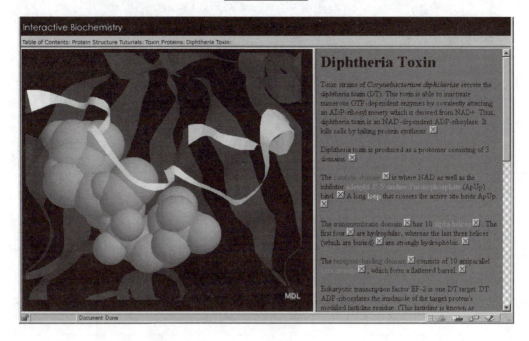

1. How does diphtheria toxin (DT) kill cells?_____
_____

2. What is ApUp, what effect does it have on diphtheria toxin, and how does it bind?____
_____

3. What property of antiparallel beta sheets makes them highly suitable for formation of beta sandwich and flattened barrel structures?_____
_____

4. What effect does DT have on elongation factor EF-2?_____
_____

5. Which domain is prompted to unfold by the protonation of a pair of His residues when the toxin enters a host cell?_____

6. The unfolding (described in question 5) continues with dissociation of a hydrophilic helix from the helical cluster. How does this lead to insertion of this domain into the bilayer?_____
_____

PTPase — A Toxin from *Yersinia*

See G&G, 2/e
p. S-29, S-30

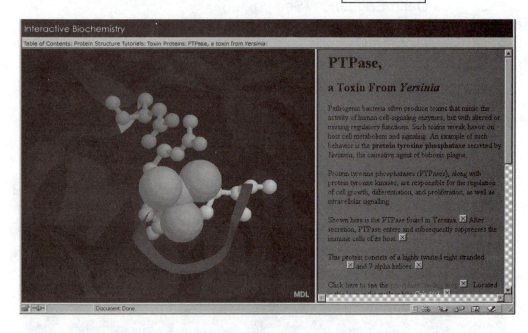

1. What is the normal function of protein tyrosine phosphatase in animal systems, and how is it related to cell signaling? Why can PTPase from *Yersinia* be harmful?_____

   _____

   _____

2. The beta sheet in this protein is twisted significantly. Use a protractor or your best judgment by eye to estimate the angle of twist from the first strand of the beta sheet to the last._____

3. Which protein residue in the phosphate-binding loop accepts the phosphoryl group during a phosphatase reaction?_____

   _____

4. Write a mechanism for the PTPase reaction, making sure to include a phosphocysteine intermediate (at Cys403), and including catalytic roles as described for Gly290, Arg409, and Asp 356.

5. Question for further reading: Look for information about PTPase 1B and other PTPase enzymes, and explore whether these enzymes use similar mechanisms and/or phosphocysteine intermediates.

Charles M. Grisham

Protein-Nucleic Acid Complexes

Aspartyl-tRNA Synthetase

See G&G, 2/e
p. 1075-1083

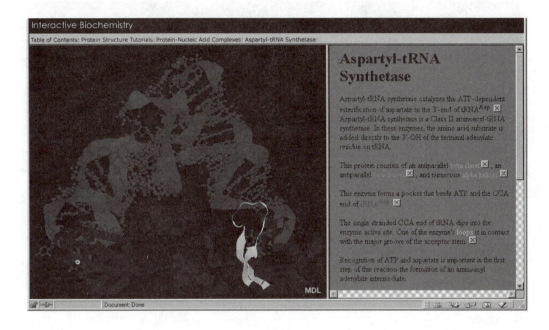

1. What is the catalytic function of aspartyl-tRNA synthetase?_____
   _____
   _____

2. Sketch partial reaction pathways below to illustrate the difference between Class I
   and Class II tRNA synthetases.

3. Near what elements of secondary structure in this enzyme are the -CCA end of tRNA
   and the ATP substrate bound?_____
   _____

4. What part of the enzyme structure is in contact with the major groove of the acceptor
   stem of tRNA?_____
   _____

5. Which element of secondary structure in the enzyme binds the anticodon loop of
   tRNA?_____

Catabolite Gene Activator Protein

See G&G, 2/e
p. 1034-1035

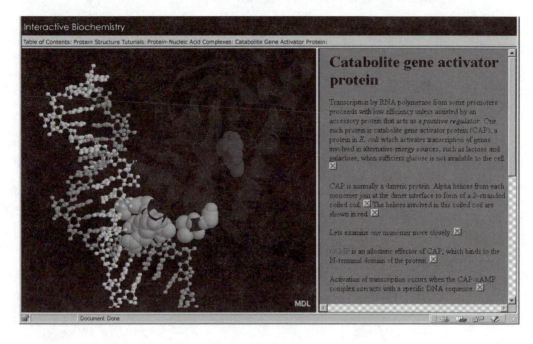

Interactive Biochemistry

Table of Contents: Protein Structure Tutorials: Protein-Nucleic Acid Complexes: Catabolite Gene Activator Protein:

Catabolite gene activator protein

Transcription by RNA polymerase from some promoters proceeds with low efficiency unless assisted by an accessory protein that acts as a *positive regulator*. One such protein is catabolite gene activator protein (CAP), a protein in *E. coli* which activates transcription of genes involved in alternative energy sources, such as lactose and galactose, when sufficient glucose is not available to the cell. ☒

CAP is normally a dimeric protein. Alpha helices from each monomer join at the dimer interface to form of a 2-stranded coiled coil. ☒ The helices involved in this coiled coil are shown in red. ☒

Lets examine one monomer more closely ☒

cAMP is an allosteric effector of CAP, which binds to the N-terminal domain of the protein ☒

Activation of transcription occurs when the CAP-cAMP complex interacts with a specific DNA sequence. ☒

MDL

Document: Done

1. Describe the function of the catabolite gene activator protein in *E. coli.*_____

    _____

    _____

2. Study this structure, and describe the formation and nature of the two-stranded coiled coil in the center of the protein's structure._____

    _____

3. What is the allosteric effector of CAP, and where does it bind in this structure? Where is it with respect to the coiled coil?_____

    _____

4. Where does DNA bind to this structure?_____

    _____

5. Does the protein interact with the major groove or the minor groove of DNA?_____

    _____

6. What structural change occurs in the DNA structure upon binding to CAP?_____

    _____

_____ Charles M. Grisham

DNA Polymerase I

See G&G, 2/e
p. 991-993

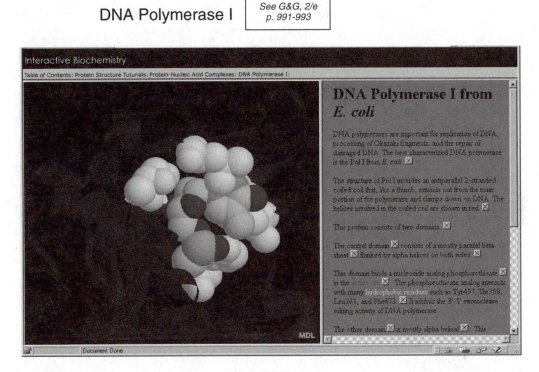

Interactive Biochemistry

Table of Contents: Protein Structure Tutorials: Protein-Nucleic Acid Complexes: DNA Polymerase I:

DNA Polymerase I from *E. coli*

DNA polymerases are important for replication of DNA, processing of Okazaki fragments, and the repair of damaged DNA. The best characterized DNA polymerase is the Pol I from *E. coli*. ☒

The structure of Pol I includes an antiparallel 2-stranded coiled coil that, like a thumb, extends out from the main portion of the polymerase and clamps down on DNA. The helices involved in the coiled coil are shown in red. ☒

This protein consists of two domains. ☒

The central domain ☒ consists of a mostly parallel beta sheet ☒ flanked by alpha helices on both sides. ☒

This domain binds a nucleoside analog phosphorothioate ☒ in the active site ☒. The phosphorothioate analog interacts with many hydrophobic residues such as Tyr497, Thr358, Leu361, and Phe473. ☒ It inhibits the 3'-5' exonuclease editing activity of DNA polymerase.

The other domain ☒ is mostly alpha helical. ☒ This

MDL

Document: Done

1. Question for further reading: DNA polymerases can be used for replicating DNA, processing Okazaki fragments, or repairing damaged DNA. What is the specific function of DNA polymerase I (Pol I) in *E. coli*?_____

_____

2. DNA polymerase I has a two-stranded coiled coil in its structure. What molecular interactions stabilize the coiled coil?_____

3. This enzyme structure contains a nucleoside phosphorothioate analog. Read further if you don't already know, and draw possibile structures for some phosphorothioates.

4. What are the hydrophobic residues that coordinate the nucleoside analog here?_____

_____

5. Which portion of the nucleoside analog interacts with the hydrophobic residues noted in question 4?_____

6. The nucleoside analog is shown in CPK colors, so that you should be able to elucidate its structure by studying the model. Study the structure (or consult the reference provided in the tutorial) and suggest a reason why it inhibits the 3',5'-exonuclease activity of Pol I._____

Elongation Factor G | See G&G, 2/e p. 1099

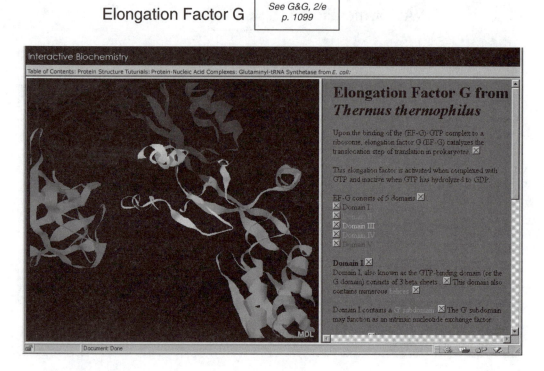

Table of Contents: Protein Structure Tutorials: Protein-Nucleic Acid Complexes: Glutaminyl-tRNA Synthetase from *E. coli:*

Elongation Factor G from *Thermus thermophilus*

Upon the binding of the (EF-G)-GTP complex to a ribosome, elongation factor G (EF-G) catalyzes the translocation step of translation in prokaryotes. ☒

This elongation factor is activated when complexed with GTP and inactive when GTP has hydrolyzed to GDP.

EF-G consists of 5 domains ☒.
☒ Domain I
☒
☒ Domain III
☒ Domain IV
☒ Domain V

Domain I ☒
Domain I, also known as the GTP-binding domain (or the G domain) consists of 3 beta sheets ☒ This domain also contains numerous helices ☒

Domain I contains a G' subdomain ☒ The G' subdomain may function as an intrinsic nucleotide exchange factor.

MDL

Document: Done

1. What controls activation and inactivation of EF-G?_____
_____

2. What types of beta sheets are found in Domain I, the GTP-binding domain?_____
_____

3. The beta barrel in Domain II is referred to as a "parallel" beta barrel, but is it? How many of the 12 strands are parallel?_____

4. Consider Domain III, both in isolation and in the native protein, and comment on the antiparallel beta sheet and on which side(s) of it you would be likely to find hydrophobic residues._____
_____

5. Where is Domain IV in the native protein? Is it exposed to solvent? Comment on the surface residues you would expect to find in this domain._____
_____

6. Again, study the location of Domain V in the native protein and comment on the nature of the surface of this domain._____

Elongation Factor TU

See G&G, 2/e
p. 1099-1100

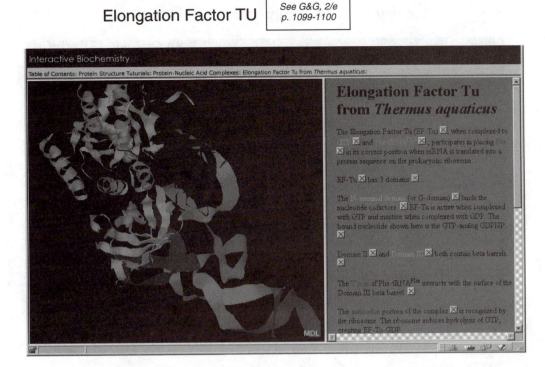

Interactive Biochemistry

Table of Contents: Protein Structure Tutorials: Protein-Nucleic Acid Complexes: Elongation Factor Tu from *Thermus aquaticus*:

Elongation Factor Tu from *Thermus aquaticus*

The Elongation Factor Tu (EF-Tu), when complexed to GTP and , participates in placing in its correct position when mRNA is translated into a protein sequence on the prokaryotic ribosome.

EF-Tu has 3 domains.

The N-terminal domain (or G-domain) binds the nucleotide cofactors. EF-Tu is active when complexed with GTP and inactive when complexed with GDP. The bound nucleotide shown here is the GTP-analog GDPNP.

Domain II and Domain III both contain beta barrels.

The T arm of Phe-tRNA^{Phe} interacts with the surface of the Domain III beta barrel.

The anticodon portion of the complex is recognized by the ribosome. The ribosome induces hydrolysis of GTP, creating EF-Tu-GDP.

1. Why is GTP required by EF-TU? How does it modulate the function of EF-TU?_____
   _____

2. "GDPNP", bound here to EF-TU, is a GTP in which the beta and gamma phosphates are joined by N rather than by O. Draw the structure of this nucleotide analog, and explain why it is ideal for forming complexes with nucleotide-hydrolyzing enzymes like this._____
   _____

3. Study the structure and the text, and describe the roles for Domain II and Domain III in the function of this protein._____
   _____
   _____

4. What inserts into the crevice between Domains I and II after GTP is hydrolyzed?_____
   _____

5. Which residues coordinate the Phe brought to this complex by the tRNA molecule?__
   _____

6. What are the roles of Switch I and Switch II?_____
   _____

Glutaminyl-tRNA Synthetase

See G&G, 2/e
p. 1075-1083

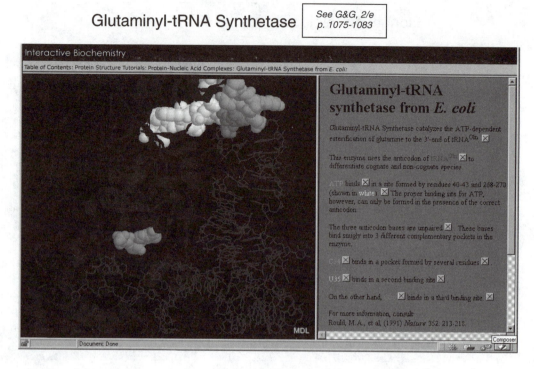

1. What is the reaction catalyzed by glutaminyl-tRNA synthetase?_____

_____

_____

2. Only binding of the correct anticodon (of tRNA) can facilitate the formation of a suitable ATP binding site on the protein. Estimate the distance between the anticodon site and the nucleotide site, using structural features you can identify in this complex.

_____

_____

3. Question for further reading: Is glutaminyl-tRNA synthetase a Class I or a Class II synthetase?_____

_____

4. Write a simple mechanism for this reaction, based on your answer to question 3 above.

Charles M. Grisham

Seryl-tRNA Synthetase

See G&G, 2/e
p. 1075-1083

Interactive Biochemistry

Table of Contents: Protein Structure Tuturials: Protein-Nucleic Acid Complexes: Seryl tRNA Synthetase:

Seryl tRNA Synthetase

Seryl tRNA synthetase specifically links serine to its cognate tRNA acceptor and thus plays an important role in protein synthesis. ☒

Shown here are two monomers of seryl tRNA synthetase. ☒ Note that the structure shown is a homodimer but looks unsymmetric. This is because residues 39-37 of one of the subunits are not visible in this structure. This segment in the other (visible) subunit forms the long coiled coil structure that will be highlighted later in this tutorial exercise.

Each monomer ☒ contains four beta sheets ☒

The N-terminal domain is made up of a long antiparallel coiled coil ☒ The N-terminal coiled coil provides the docking site for tRNA ☒

Upon tRNA binding, the coiled coil rotates by approximately 23 degrees, thereby allowing it to fit inside a cleft between the variable arm and the D loop. This helical arm then directs the tRNA acceptor stem into the active site of the ☒

For more information consult:

1. Here's a lesson in X-ray crystallography. The model shown is a homodimer. However, it appears to be distinctly asymmetric. How can this be? As stated, part of the molecule is not visible in this structural model. How can this happen? (You may need to do additional reading to grasp this point.)_____
_____
_____
_____

2. Even though the beta sheets are not labeled with arrows in this model, study the molecule and determine whether the beta sheets in the monomer shown are parallel or antiparallel._____
_____
_____

3. What is the purpose of the N-terminal coiled coil?_____
_____
_____

4. What change occurs in the coiled coil domain upon tRNA binding?_____
_____
_____

Saunders Interactive Biochemistry Manual and Workbook _____ 95

TATA Box and TATA-Binding Proteins

See G&G, 2/e
p. 1028-1029

Interactive Biochemistry

Table of Contents: Protein Structure Tutorials: Protein-Nucleic Acid Complexes: The TATA Box and TATA-Binding Protein:

The TATA Box and TATA-Binding Protein

The TATA module is a short sequence of DNA, which indicates the initiation site and defines the promoter of protein-encoding genes ☒. The consensus sequence of the TATA box is "A AAAA ☒

These base pairs of the TATA box bind the underside of a TATA Binding Protein ☒

TATA Binding Proteins (TBPs) are regulatory proteins that are necessary for activation of transcription ☒. These proteins are made up of a ten-stranded ☒ and four ☒.

TBP sits on top of the TATA box just as a saddle sits on a horse ☒ The bound DNA base pairs bend severely toward the major groove, significantly altering the trajectory of the DNA by creating a 100 degree bend in the axis of the DNA. ☒

The bound TBP pries open the minor groove, forcing the TATA sequence to unwind, and thereby creating a mostly hydrophobic interface with the underside of the TATA

MDL

Document: Done

1. What is a TATA box, and what is the consensus sequence found in it?_____
   _____
   _____

2. Describe the properties and function of the TATA binding protein._____
   _____
   _____

3. Is the beta sheet in the TATA binding protein parallel or antiparallel? Where would
   you expect to find hydrophobic residues around this sheet?_____
   _____
   _____

4. What molecular forces dictate the interaction of the beta sheet with the DNA
   segment? What does the TATA binding protein do to the DNA structure to expose
   appropriate groups on the DNA?_____
   _____
   _____

_____ Charles M. Grisham

Cell Signaling and Hormones

Signaling Domains

See G&G, 2/e
p. S-32 - S-35

PDZ Domains

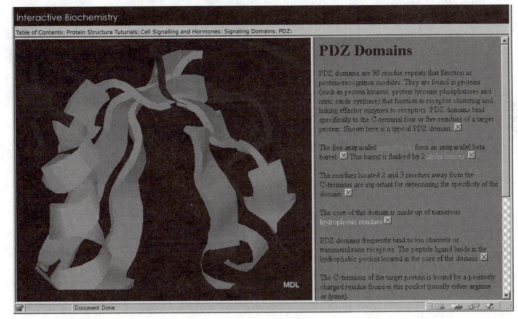

Interactive Biochemistry

Table of Contents: Protein Structure Tutorials: Cell Signalling and Hormones: Signaling Domains: PDZ:

PDZ Domains

PDZ domains are 90 residue repeats that function as protein-recognition modules. They are found in proteins (such as protein kinases, protein tyrosine phosphatases and nitric oxide synthase) that function in receptor clustering and linking effector enzymes to receptors. PDZ domains bind specifically to the C-terminal four or five residues of a target protein. Shown here is a typical PDZ domain. ⊠

The five antiparallel .... form an antiparallel beta barrel ⊠ This barrel is flanked by 2 alpha helices ⊠

The residues located 2 and 3 residues away from the C-terminus are important for determining the specificity of the domain. ⊠

The core of this domain is made up of numerous hydrophobic residues ⊠

PDZ domains frequently bind to ion channels or transmembrane receptors. The peptide ligand binds in the hydrophobic pocket located in the core of the domain. ⊠

The C-terminus of the target protein is bound by a positively charged residue found in this pocket (usually either arginine or lysine).

MDL

Document: Done

1. In what kinds of proteins are PDZ domains commonly found?_____
   _____

2. What is the protein motif that the PDZ domain recognizes and binds?_____
   _____

3. Where are the alpha helices located with respect to the beta barrel in this domain?____
   _____

4. Which residues on the PDZ domain influence binding specificity?_____
   _____

5. Examine the core of this domain and suggest a driving force for folding of this
   domain._____
   _____

6. Target peptides are thought to bind to PDZ domains by insertion into the hydrophobic
   core of the domain. What residue in this pocket coordinates (and neutralizes) the C-
   terminal carboxyl group of target peptides?_____

PH Domains

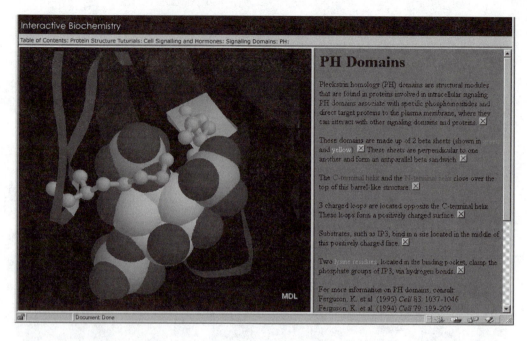

PH Domains

Pleckstrin homology (PH) domains are structural modules that are found in proteins involved in intracellular signaling PH domains associate with specific phosphoinositides and direct target proteins to the plasma membrane, where they can interact with other signaling domains and proteins. ☒

These domains are made up of 2 beta sheets (shown in and yellow) ☒ These sheets are perpendicular to one another and form an antiparallel beta sandwich ☒

The C-terminal helix and the N-terminal helix close over the top of this barrel-like structure. ☒

3 charged loops are located opposite the C-terminal helix These loops form a positively charged surface ☒

Substrates, such as IP3, bind in a site located in the middle of this positively charged face. ☒

Two lysine residues, located in the binding pocket, clamp the phosphate groups of IP3, via hydrogen bonds. ☒

For more information on PH domains, consult:
Ferguson, K. et al (1995) *Cell* 83: 1037-1046
Ferguson, K. et al. (1994) *Cell* 79: 199-209

MDL

Document: Done

1. To what do PH domains bind, and how do they function?_____
   _____

2. The core of the PH domain is an orthogonal pair of antiparallel beta sheets. In this respect the PH domain is similar to what other signaling domain?_____

3. Where are the N-terminal and C-terminal alpha helices located in this domain?_____
   _____

4. Where are the three charged loops located on this domain?_____
   _____

5. What is the apparent function of the three charged loops?_____
   _____

6. Where are the lysine amino groups of the positively charged loops situated?_____
   _____

7. What would you expect to be the ionization state of these lysine residues?_____
   _____

PTB Domains

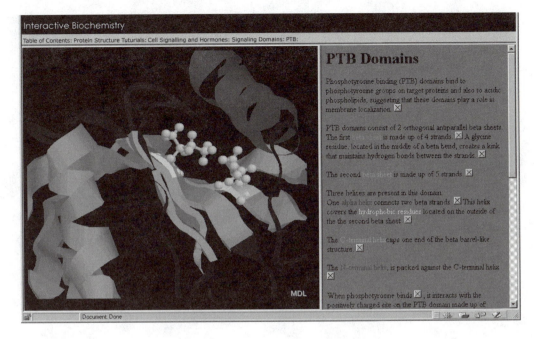

PTB Domains

Phosphotyrosine binding (PTB) domains bind to phosphotyrosine groups on target proteins and also to acidic phospholipids, suggesting that these domains play a role in membrane localization. ☒

PTB domains consist of 2 orthogonal antiparallel beta sheets. The first _____ is made up of 4 strands. ☒ A glycine residue, located in the middle of a beta bend, creates a kink that maintains hydrogen bonds between the strands. ☒

The second _____ is made up of 5 strands. ☒

Three helices are present in this domain.
One alpha helix connects two beta strands ☒ This helix covers the hydrophobic residues located on the outside of the the second beta sheet. ☒

The C-terminal helix caps one end of the beta barrel-like structure. ☒

The N-terminal helix is packed against the C-terminal helix. ☒

When phosphotyrosine binds ☒ , it interacts with the positively charged site on the PTB domain made up of

MDL

Document: Done

1. Why is the PTB domain postulated to be involved in membrane localization of its target proteins?_____
   _____

2. A glycine residue in a beta bend helps to maintain hydrogen bonding in the first of two beta sheets in this protein. What makes glycine an appropriate residue for beta bends?_____
   _____

3. What are the apparent functions of the three alpha helices in this domain?_____
   _____
   _____
   _____

4. What four residues in this domain coordinate bound phosphotyrosine?_____
   _____
   _____

5. PTB target peptides contain a hydrophobic residue five residues from the phosphotyrosine group. How does this hydrophobic residue bind to the PTB domain?
   _____
   _____
   _____

SH2 Domains

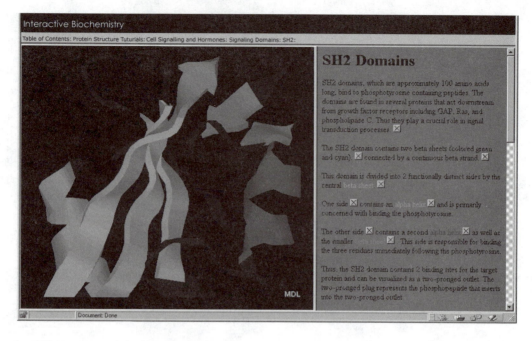

SH2 Domains

SH2 domains, which are approximately 100 amino acids long, bind to phosphotyrosine containing peptides. The domains are found in several proteins that act downstream from growth factor receptors including GAP, Ras, and phospholipase C. Thus they play a crucial role in signal transduction processes. ☒

The SH2 domain contains two beta sheets (colored green and cyan), ☒ connected by a continuous beta strand. ☒

This domain is divided into 2 functionally distinct sides by the central beta sheet. ☒

One side ☒ contains an alpha helix ☒ and is primarily concerned with binding the phosphotyrosine.

The other side ☒ contains a second alpha helix ☒ as well as the smaller ☒ This side is responsible for binding the three residues immediately following the phosphotyrosine.

Thus, the SH2 domain contains 2 binding sites for the target protein and can be visualized as a two-pronged outlet. The two-pronged plug represents the phosphopepide that inserts into the two-pronged outlet.

MDL

Document: Done

1. What are some signaling proteins that are know to contain SH2 domains?_____
   _____

2. What is unusual about the two beta sheets found in this protein?_____
   _____

3. Describe the two faces of this signaling domain (on either side of the beta sheet), both
   in terms of structure and in terms of function._____
   _____
   _____

4. Describe the residues that form the phosphotyrosine binding site._____
   _____

5. Describe the "induced-fit" response that occurs upon phosphotyrosine binding,
   including the role of Arg32._____
   _____
   _____

6. What elements of domain structure confer peptide binding specificity to this domain?
   _____

7. What is the function of Lys57?_____
   _____

SH3 Domains

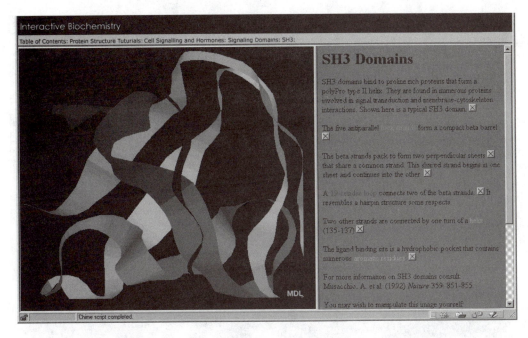

SH3 Domains

SH3 domains bind to proline rich proteins that form a polyPro type II helix. They are found in numerous proteins involved in signal transduction and membrane-cytoskeleton interactions. Shown here is a typical SH3 domain. ☒

The five antiparallel _____ form a compact beta barrel ☒

The beta strands pack to form two perpendicular sheets ☒ that share a common strand. This shared strand begins in one sheet and continues into the other. ☒

A 19-residue loop connects two of the beta strands. ☒ It resembles a hairpin structure some respects.

Two other strands are connected by one turn of a _____ (135-137) ☒

The ligand binding site is a hydrophobic pocket that contains numerous _____ residues ☒

For more information on SH3 domains consult Musacchio, A. et al. (1992) *Nature* 359: 851-855.

You may wish to manipulate this image yourself

Chime script completed.

1. What structural motifs in target proteins are recognized and bound by SH3 domains?
    _____

2. This domain can be viewed with as a beta barrel or as a pair of beta sheets that share a common strand. Taking the latter view, what other signaling domain is similar to SH3?_____

3. Where is the hairpin loop located in this signaling domain?_____

4. Which two beta strands (numbered according to the order of their appearance in the peptide sequence) are joined by a 1-turn alpha helix?_____

5. What is the nature of the ligand binding site on this domain?_____

6. What other signaling domain binds to proline-rich domains?_____

WW Domains

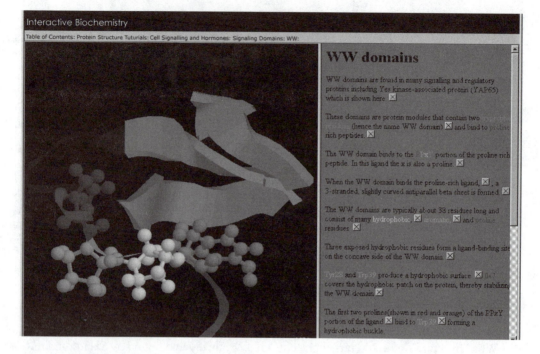

WW domains

WW domains are found in many signalling and regulatory proteins including Yes kinase-associated protein (YAP65) which is shown here.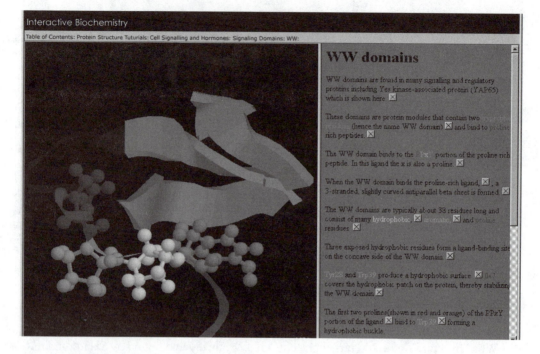

These domains are protein modules that contain two _____ (hence the name WW domain) 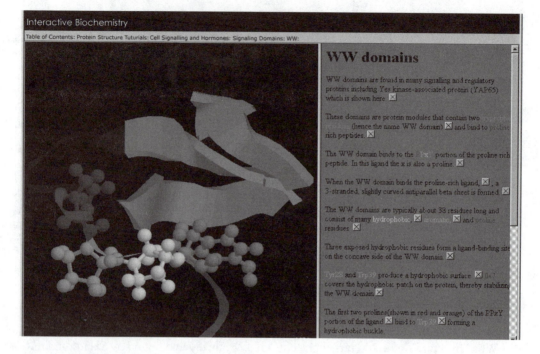 and bind to proline rich peptides.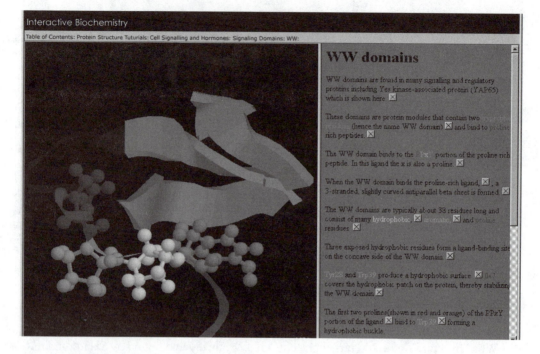

The WW domain binds to the _____ portion of the proline rich peptide. In this ligand the x is also a proline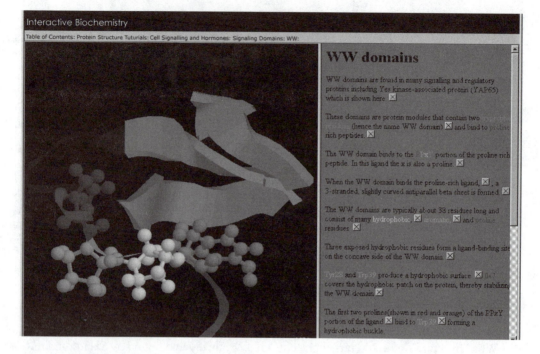

When the WW domain binds the proline-rich ligand, 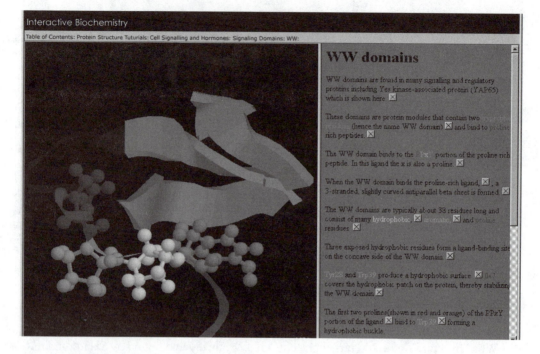, a 3-stranded, slightly curved antiparallel beta sheet is formed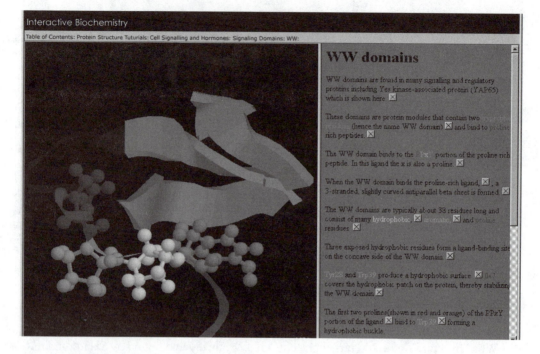

The WW domains are typically about 38 residues long and consist of many hydrophobic 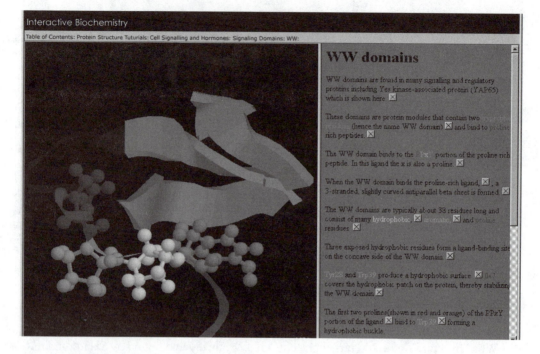 aromatic and proline residues.

Three exposed hydrophobic residues form a ligand-binding site on the concave side of the WW domain.

Tyr28 and Trp39 produce a hydrophobic surface Ile7 covers the hydrophobic patch on the protein, thereby stabilizing the WW domain

The first two prolines(shown in red and orange) of the PPxY portion of the ligand bind to Trp39 forming a hydrophobic buckle.

1. What kinds of peptide ligands bind to WW domains, and what is the consensus sequence required?_____

2. What structural feature is found in all WW domains (and gives them their name)?____
_____

3. What are the functions of Tyr28, Trp39, and Ile7 of the WW domain?_____
_____

4. Where is the ligand binding site located on the WW domain?_____
_____

5. What is the "hydrophobic buckle" formed between the WW domain and its ligand?___
_____

6. When a ligand contains a third consecutive proline residue, how is it coordinated to the WW domain?_____

7. How is the tyrosine residue of the PPXY ligand motif coordinated by the WW domain?_____
_____

Adenylyl Cyclase

See G&G, 2/e
p. S-6

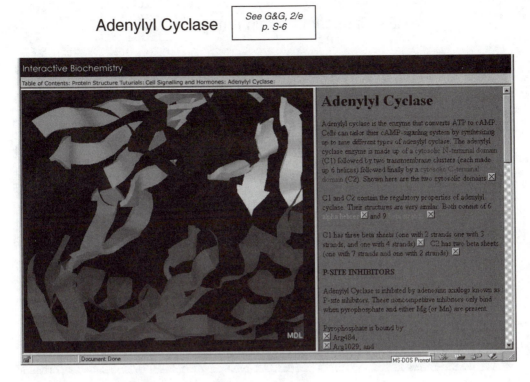

Interactive Biochemistry

Table of Contents: Protein Structure Tutorials: Cell Signalling and Hormones: Adenylyl Cyclase:

Adenylyl Cyclase

Adenylyl cyclase is the enzyme that converts ATP to cAMP. Cells can tailor their cAMP-signaling system by synthesizing up to nine different types of adenylyl cyclase. The adenylyl cyclase enzyme is made up of a cytosolic N-terminal domain (C1) followed by two transmembrane clusters (each made up 6 helices) followed finally by a cytosolic C-terminal domain (C2). Shown here are the two cytosolic domains ☒

C1 and C2 contain the regulatory properties of adenylyl cyclase. Their structures are very similar. Both consist of 6 alpha helices ☒ and 9 ☒

C1 has three beta sheets (one with 2 strands one with 3 strands, and one with 4 strands) ☒ C2 has two beta sheets (one with 7 strands and one with 2 strands) ☒

P-SITE INHIBITORS

Adenylyl Cyclase is inhibited by adenosine analogs known as P-site inhibitors. These noncompetitive inhibitors only bind when pyrophosphate and either Mg (or Mn) are present.

Pyrophosphate is bound by
☒ Arg484,
☒ Arg1029, and

MDL

Document: Done   MS-DOS Prompt

1. What are P-site inhibitors, and what are their properties?_____
    _____

2. What amino acid residues constitute the P-site inhibitor binding site?_____
    _____

3. What are the roles of Arg1029 and Asn1025?_____
    _____

4. Identify the residues from each domain that participate in forskolin coordination._____
    _____

5. How does the G protein affect adenylyl cyclase activity?_____
    _____

6. Describe at the molecular level the interactions of the G protein with adenylyl
    cyclase._____
    _____
    _____

cAMP-Dependent Protein Kinase

See G&G, 2/e
p. S-10

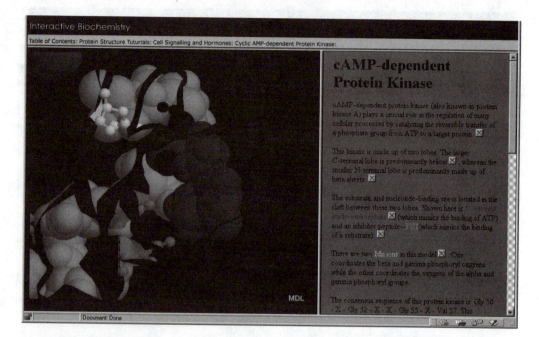

1. Where do the ATP and target peptide substrates bind to cAMP-dependent protein kinase A (PKA)?_____

   _____

2. What is the "consensus sequence" found at the nucleotide-binding site in this and other similar kinases?_____

   _____

3. Which amino acid residues are believed to be important in the phosphoryl transfer reaction?_____

4. Which amino acid residue at the active site coordinates and stabilizes the gamma phosphoryl group of ATP?_____

5. Which residue is thought to serve as a catalytic base in the phosphoryl transfer reaction?_____

6. Which residues are thought to be involved in substrate peptide recognition?_____

   _____

Charles M. Grisham

Ca-Myristoyl Switches (See page 61)

Calmodulin

See G&G, 2/e
p. S-20

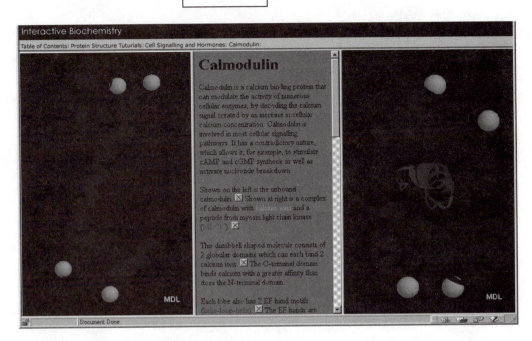

1. What is the function of calmodulin in biological systems?_____
   _____
   _____

2. Which of the two domains of calmodulin — N-terminal or C-terminal — binds Ca^{2+}
   ions more tightly?_____

3. Explain how the small antiparallel beta sheet that joins the pair of EF hands in each of
   the globular domains of calmodulin is actually intertwined with the "hands"._____
   _____
   _____

4. What change does Ca^{2+} binding evoke for the hydrophobic patches on calmodulin?___
   _____

5. What do all calmodulin-binding proteins possess?_____

6. What becomes of the long central helix of calmodulin when it forms a complex with a
   target peptide or calmodulin-binding protein?_____
   _____

7. Which residues of the long central helix mediate this change?_____

_____

Cholera Toxin (See page 85)
Cyclooxygenase (See page 75)

Farnesyl Transferase   See G&G, 2/e
                         p. 278

1. Describe the nature of the protein substrate for a farnesyl transferase reaction._____
_____

2. How many alpha helices can you find in the crescent-shaped alpha subunit? How
   many coiled coils?_____

3. The beta subunit assembles 12 helices into a novel three-dimensional structure.
   Describe._____
   _____

4. Where do the protein substrate and farnesyl pyrophosphate bind on the FTase?_____
   _____

5. What are the amino acid residues that bind the protein substrate?_____
   _____

6. What residues bind the farnesyl pyrophosphate?_____

7. Where does the Cys residue that is the farnesylation site bind on the FTase?_____
   _____

FAS Death Domain

See G&G, 2/e p. S-26

Fas Death Domain

Apoptosis, another name for programmed cell death, is a process crucial to embryogenesis, metamorphosis, immune system function, and normal cell turnover in higher organisms. Apoptosis is initiated in a variety of ways. One of these involves "death factor" proteins such as **fas** and **tumor necrosis factor**. Fas is a cytokine receptor which mediates apoptosis. The region of fas involved in signaling apoptosis is called the death domain. Residues 202-319 make up this domain. ☒

This domain consists of 6 amphipathic, antiparallel alpha helices. ☒ Helices 1 and 2 are found in the center of the domain, ☒ while 3 and 4 are on one side ☒ and 5 and 6 on the opposite side. ☒ A crossing pattern is formed by a which connects helices 4 and 5 while bisecting helices 1 and 2. ☒

The domain's ability to self associate is mediated by helices 2 and 3 ☒

Helices 5 and 6 play an important role in the binding of allosteric effectors ☒ The hydrophobic site formed by these two helices is a potential binding site. Hydrophobic residues are shown in yellow ☒

MDL

Document Done

1. What is apoptosis?_____
   _____

2. Which residues of the Fas protein make up the "death domain", and why is it called
   that?_____
   _____

3. The cluster of alpha helices in this domain is unusual. Describe it._____
   _____
   _____

4. Which of the helices mediate self-association between death domains?_____

5. What do helices 5 and 6 enable for this domain?_____
   _____

6. Question for further research: The fas death domain (shown here) associates with
   other similar domains on proteins that contain "death effector" domains, which bind
   to "caspases", a family of proteases linked to apoptosis. What are caspases, and what
   are the events by which they trigger apoptosis?

Charles M. Grisham

G Proteins

See G&G, 2/e p. S-4 – S-10

Interactive Biochemistry

Table of Contents: Protein Structure Tutorials: Cell Signalling and Hormones: G Protein:

The Heterotrimeric G Protein Complex

A Molecular Nanomachine

Hundreds of cell-surface receptors for hormones and other ligands use 'G proteins' to transduce intracellular signalling pathways. Dramatic recent advances by several laboratories have provided exciting details of the structure of the heterotrimeric G proteins ⊠ With mating receptor proteins that act like molecular levers, a trio of 'switch domains' that can adopt 'open' or 'closed' conformations, and a beta subunit that resembles a propeller, these proteins may be thought of as molecular 'nanomachines' ⊠

The typical heterotrimeric G protein consists of an alpha subunit of 45-47 kD, ⊠ a beta subunit of 35 kD, ⊠ and a gamma subunit of 7-9 kD ⊠

The Alpha Subunit ⊠

The alpha subunit of heterotrimeric G proteins consists of two domains, a GTPase domain and an alpha-helical domain. The GTPase domain ⊠ is similar in structure to p21 ras and

MDL

Document: Done

1. Describe the subunits of a heterotrimeric G protein (and their respective MWs)._____

   _____

2. Describe the "linkers" in the alpha subunit and what binds between them._____

   _____

3. Describe Switch I, Switch II, and Switch III in the alpha subunit, and explain why
   they are important._____

   _____

4. Why has nature designed the beta subunit as a seven-bladed propeller?_____

   _____

5. What is a WD motif, and how are the WD motifs of the beta subunit arranged with
   respect to the "propeller blades"?_____

   _____

6. Describe the lipid anchors that are found in these heterotrimeric G proteins._____

   _____

GAP (GTPase Activating Protein)

See G&G, 2/e p. S-10

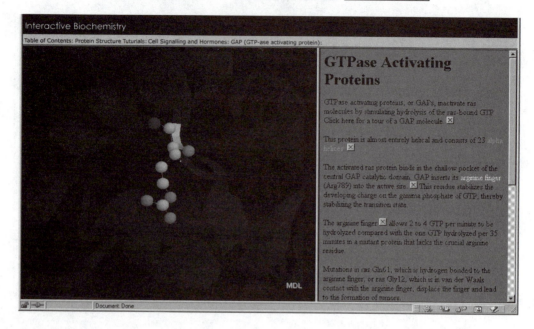

GTPase Activating Proteins

GTPase activating proteins, or GAPs, inactivate ras molecules by stimulating hydrolysis of the ras-bound GTP. Click here for a tour of a GAP molecule. ☒

This protein is almost entirely helical and consists of 23 alpha helices. ☒

The activated ras protein binds in the shallow pocket of the central GAP catalytic domain. GAP inserts its arginine finger (Arg789) into the active site. ☒ This residue stabilizes the developing charge on the gamma phosphate of GTP, thereby stabilizing the transition state.

The arginine finger ☒ allows 2 to 4 GTP per minute to be hydrolyzed compared with the one GTP hydrolyzed per 35 minutes in a mutant protein that lacks the crucial arginine residue.

Mutations in ras Gln61, which is hydrogen bonded to the arginine finger, or ras Gly12, which is in van der Waals contact with the arginine finger, displace the finger and lead to the formation of tumors.

MDL

Document Done

---

1. Explain the activity of "GAPs" in terms of how ras proteins function._____
   _____

2. How many separate alpha helices can you find in this protein?_____

3. What is the "arginine finger" of the GAP, and how does it function?_____
   _____

4. What is the rate of GTP hydrolysis on ras in the presence and absence of GAP?_____
   _____

5. Describe the mutations related to the arginine finger on GAP and on ras that can lead
   to tumor formation._____
   _____

MAP Kinase

See G&G, 2/e
p. S-34

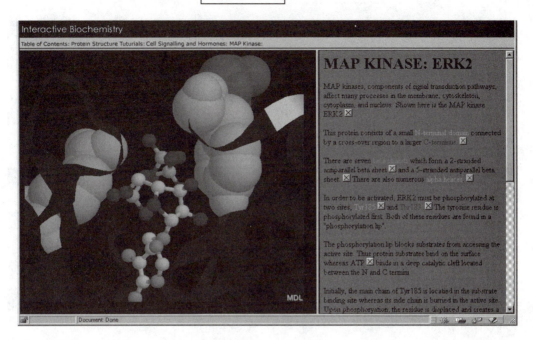

Interactive Biochemistry

Table of Contents: Protein Structure Tutorials: Cell Signalling and Hormones: MAP Kinase:

MAP KINASE: ERK2

MAP kinases, components of signal transduction pathways, affect many processes in the membrane, cytoskeleton, cytoplasm, and nucleus. Shown here is the MAP kinase ERK2 ☒

This protein consists of a small N-terminal domain connected by a cross-over region to a larger C-terminus. ☒

There are seven ☐ which form a 2-stranded antiparallel beta sheet ☒ and a 5-stranded antiparallel beta sheet. ☒ There are also numerous alpha helices ☒

In order to be activated, ERK2 must be phosphorylated at two sites, Tyr185 ☒ and Thr183 ☒ The tyrosine residue is phosphorylated first. Both of these residues are found in a "phosphorylation lip".

The phosphorylation lip blocks substrates from accessing the active site. Thus protein substrates bind on the surface whereas ATP ☒ binds in a deep catalytic cleft located between the N and C termini.

Initially, the main chain of Tyr185 is located in the substrate binding site whereas its side chain is buried in the active site. Upon phosphorylation, the residue is displaced and creates a

Document: Done

MDL

1. Question for additional research: What does MAP refer to in the MAP kinases?_____
   _____

2. Describe the beta sheets found in this MAP kinase._____
   _____

3. Phosphorylation at two sites is required to activate the MAP kinase. What are those sites, and in what order must they be phosphorylated?_____
   _____

4. Where does ATP bind in the active site, relative to the phosphorylation lip?_____
   _____

5. Which residues interact with and stabilize the phosphate on Tyr185?_____
   _____

6. What residue interacts with and stabilizes the phosphate on Thr183?_____
   _____

7. What is the consequence for activity of removal of either of these phosphates?_____
   _____

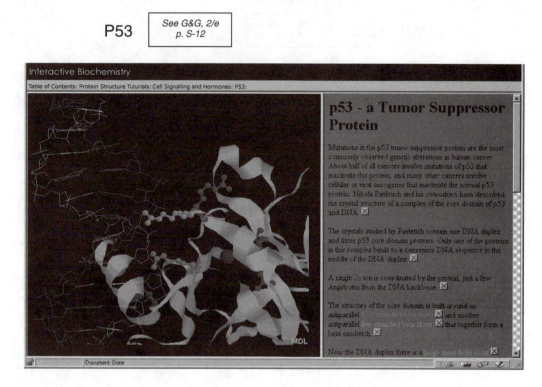

1. The p53 protein contains a single zinc ion. Where is it located?_____

_____

2. Describe the structure of the core domain._____

_____

3. Describe the segments of the protein and the residues that surround the zinc site._____

_____

4. What is meant by the term "mutation hot spots"?_____

_____

5. Suggest reasonable explanations for the tumorigenic potential of mutations at Gly-245, Arg-248, and Arg-273. In other words, insofar as you can determine from the structure shown, why are these residues important for the proper function of p53, and why might tumors form if these residues are mutated?_____

_____

_____

_____

Protein Kinase C

See G&G, 2/e
p. S-21 – S-23

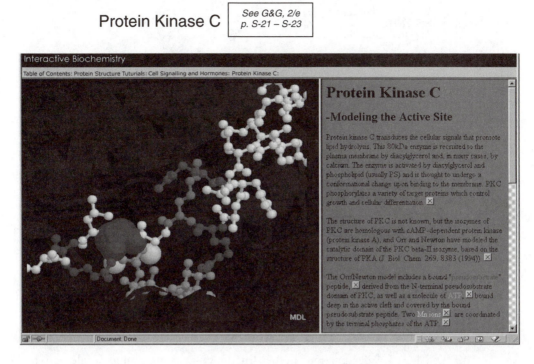

Interactive Biochemistry

Table of Contents: Protein Structure Tutorials: Cell Signalling and Hormones: Protein Kinase C:

Protein Kinase C

-Modeling the Active Site

Protein kinase C transduces the cellular signals that promote lipid hydrolysis. This 80kDa enzyme is recruited to the plasma membrane by diacylglycerol and, in many cases, by calcium. The enzyme is activated by diacylglycerol and phospholipid (usually PS) and is thought to undergo a conformational change upon binding to the membrane. PKC phosphorylates a variety of target proteins which control growth and cellular differentiation.

The structure of PKC is not known, but the isozymes of PKC are homologous with cAMP-dependent protein kinase (protein kinase A), and Orr and Newton have modeled the catalytic domain of the PKC beta-II isozyme, based on the structure of PKA (J. Biol. Chem. 269, 8383 (1994)).

The Orr/Newton model includes a bound "pseudosubstrate" peptide, derived from the N-terminal pseudosubstrate domain of PKC, as well as a molecule of ATP, bound deep in the active cleft and covered by the bound pseudosubstrate peptide. Two Mn ions are coordinated by the terminal phosphates of the ATP.

MDL

Document Done

1. How was this postulated structure for protein kinase C (PKC) created?_____
   _____

2. Question for additional reading: What is a pseudosubstrate domain with respect to
   PKC?_____
   _____

3. What kind of beta sheet is found in this modeled structure? Is its location in the
   structure consistent with what you know about such beta sheets?_____
   _____

4. What is the postulated site of autophosphorylation, and where is it located?_____
   _____

5. Why would the pseudosubstrate peptide shown here not be subject to phosphorylation
   by PKC?_____

6. What are the distinguishing and essential residues of the pseudosubstrate peptide?____
   _____

PTPase — A Toxin from *Yersinia* (See page 88)
Ras: a GTPase Enzyme (See page 66)

The Sweet Isomers

See G&G, 2/e
p. 209-217

Introduction

Learning the structures of the simple sugars presents a formidable challenge to biochemistry students. The task appears simple at first. For example, aldohexoses all possess an aldehyde function at C-1, -CHOH groups at positions 2, 3, 4, and 5, and a CH_2OH at C-6. However, with four chiral centers, 16 possible aldohexose structures can be drawn.

This set of exercises provides an interactive opportunity to identify and learn the many structures of the simple sugars. Identification exercises are provided for the aldotetroses, aldopentoses, aldohexoses, ketopentoses, and ketohexoses. In each of these, selecting a sugar from the list provided displays its structure on the screen, together with its traditional name.

Each of these structures is "live" and clickable. Clicking on any of these chiral carbons in the structure will cause the configuration at that position to invert, with an appropriate change of name. This feature can be used to compare structures and explore structural relationships in a variety of ways.

The "D-" and "L-" designations in the structural names offer another interactive feature of these exercises. Clicking on "D-" or "L-"will change the display to the opposite enantiomer of the displayed molecule. Thus, clicking on "D-" of D-glucose will invert ALL of the chiral centers at once to display L-Glucose. Clicking on the "L-" of L-ribose will change all of the chiral centers to display D-ribose.

Charles M. Grisham

Note that the "icon" on the left side of the window shows boxes around the carbon atoms that are chiral in each group of sugars. Configuration can only invert at these indicated positions. Note also that the image area on the right is blank when you begin any of these exercises. You must choose a sugar from the drop-down menu in order to display a sugar and begin the exercise.

Each set of exercises presents: (a) linear structures as Fischer ` rojections and (b) cyclic structures as Haworth ` rojections. In any of the sections of this exercise, both (a) and (b) offer the same family of sugar structures for study.

Learn the Structures and Stereoisomer Comparisons

Once you have chosen one of the classes of sugar structures (Aldohexoses, for example), and chosen either Fischer or Haworth presentation, you may also choose either of two exercises. **Learn the Structures**, as its name implies, is directed to helping you learn the structures of the simple sugars. The appearance of the screen for these exercises is like the one shown on the previous page of this manual. Choose a sugar from the drop-down menu, and its image will be displayed on the right of your screen. You may click on any of the chiral centers to invert configuration at that position only, or you may click on "D-" or "L-" to invert configuration at all of the chiral centers simultaneously.

If you choose the other type of exercise, **Stereoisomer Comparisons**, you will be presented with a slightly more complex display. In these exercises, there are two drop-down menus and two image-viewing areas on the display:

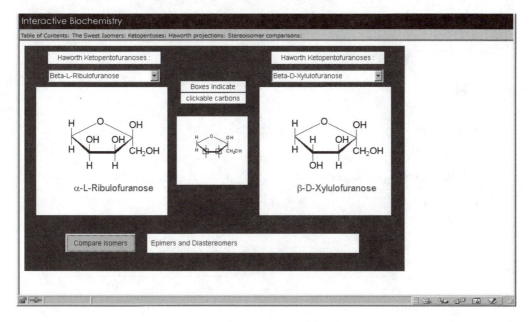

Choose a sugar from each of the drop-down menus, and they both will be displayed on the screen. Clicking on the "Compare Isomers" button at the bottom left of the screen produces a message in the white box to the right of the button, indicating the stereoisomeric relationships between the two structures you have chosen. Once again, the chiral carbons AND the "D-" and "L-" designations are live and clickable, allowing you to quickly and easily change the displayed molecules. (The icon in the middle of the display in each case shows you the carbons that are chiral and thus clickable.) Each time that you change one or both of the displayed molecules, study the structures, try to decide what the stereoisomeric relationships are, and then click on "Compare Isomers" to check your conclusion.

Questions:

1. Write the correct names for each of these structures in the blanks.

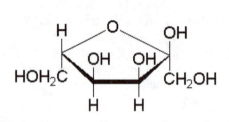

Name:_____

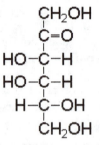

Name:_____

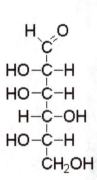

Name:_____

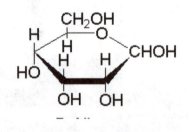

Name:_____

Charles M. Grisham

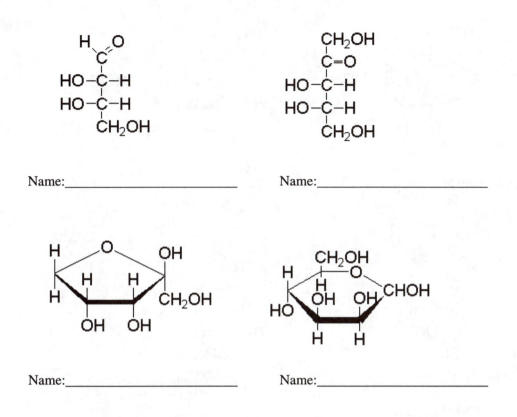

Name:_____   Name:_____

Name:_____   Name:_____

2. For each of the pairs of sugars shown below, indicate the stereochemical
    relationships.

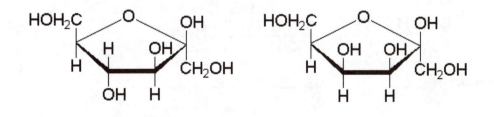

Relationship(s):_____

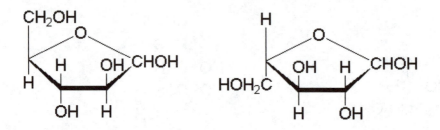

Relationship(s):_____

Relationship(s):_____

Relationship(s):_____

Charles M. Grisham

Phospholipid Structure

See G&G, 2/e
p. 243-247

A glycerophospholipid is a 1,2-diacylglycerol that has a phosphate group esterified at carbon atom 3 of the glycerol backbone. Also known as phosphoglycerides or glycerol phosphatides, these lipids form one of the largest classes of natural lipids and one of the most important. They are essential components of cell membranes and are formed in small concentrations in other parts of cells. Breakdown products of the glycerophospholipids are important signaling molecules in all animal cells.

Learning the structures and names of the glycerophospholipids can be a challenging exercise. This Java applet is intended to make it easier. It allows you to construct a glycerophospholipid and study both the structure and the name of the molecule you have created. When you begin this applet, the display will look like this:

The display features list boxes from which you can select fatty acid chains and also a head group for the glycerophospholipid you are constructing.

Questions:

1. What is the name for the glycerophospholipid that contains stearic acid at the 1-position, linoleic acid at the 2-position, and ethanolamine as the head group?_____

_____

2. What is the name for the glycerophospholipid that contains palmitic acid at the 1-position, arachidonic acid at the 2-position, and choline as the head group?_____

_____

3. Which of the following glycerophospholipids would be more commonly found in animal cells: 1-stearoyl-2-linolenoyl-phosphatidylserine or 1-linolenoyl-2-stearoyl-phosphatidylserine? Explain your answer._____

_____

4. The glycerol phosphate in natural phosphoglycerides is named *sn*-glycerol-3-phosphate. Explain this nomenclature._____

_____

5. Question for further reading: Describe all the possible breakdown products of 1-stearoyl-2-arachidonoyl-phosphatidylinositol that could function as signaling molecules in a cell._____

_____

_____

Charles M. Grisham

Structure of Nitrogenous Bases and Nucleosides

See G&G, 2/e
p. 327-334

Introduction

Familiarity with the names and structures of the purine and pyrimidine bases, and the nucleosides and nucleotides derived from them, is a requisite skill for any student of biochemistry. At the same time, it is not easy to know just when you have fully mastered these structures, and the naming conventions can seem confusing at first. For example, cytosine, with an "osine" suffix is a pyridine base, but adenosine and guanosine, with the same "osine" suffix, are purine nucleosides.

This exercise provides an easy opportunity to learn the names and structures of the nitrogenous bases and their nucleosides. It displays structures of bases and nucleosides in random order, and asks the user to identify them. When the displayed molecule is correctly identified, a new one can be displayed, and so on. This Java applet also keeps track of the user's responses, providing cumulative scoring for each session.

Name the Purines, Pyrimidines, and Nucleosides

When the applet is started, a screen similar to this one appears:

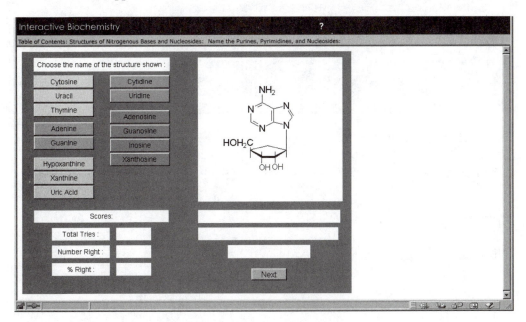

On the upper right side of the applet, you will see an image field that displays a nitrogenous base or nucleoside. On the upper left side are groups of clickable buttons named for the common nitrogenous bases and nucleosides. The fields on the lower left are for scoring, and the fields in the lower right will provide responses as you work through the exercise.

Study the displayed molecule, and select and click the button bearing its name. If your selection is correct, the name you have chosen will appear in the text field under the displayed molecule, and the applet will acknowledge your choice as "Correct!" You may then click the "Next" button in the lower right of the screen to proceed to another case. Note that the scoring area in the lower left will have recorded the total number of tries (clicks) you have made, the number right, and the percentage of your total clicks that have been correct. If you correctly identified the first displayed model with a single click, the screen will look like this:

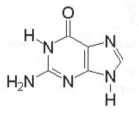

If you continue through this exercise, new molecules will be displayed every time you click "Next". The applet will keep track of your cumulative score as you proceed through the exercise. The purines, pyrimidines, and nucleosides should soon be familiar to you.

Questions:

Write the names of the structures in the blanks.

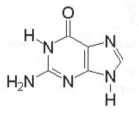

Name:_____

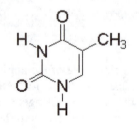

Name:_____

Charles M. Grisham

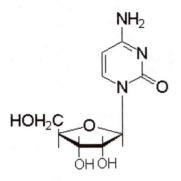

Name:_____

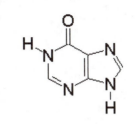

Name:_____

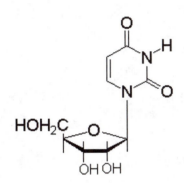

Name:_____

Name:_____

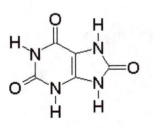

Name:_____

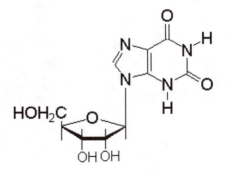

Name:_____

3. Draw the structures of the following.

Adenosine: Inosine:

Uracil: Cytosine:

Adenine: Xanthine:

Uric acid: Hypoxanthine:

 Charles M. Grisham

The Restrictions Sites of Plasmids and Genes

See G&G, 2/e
p. 350-353, 396-398

Restriction endonucleases are enzymes, isolated chiefly from bacteria, that have the ability to cleave double-stranded DNA. Plasmids are naturally occurring, circular, extrachromosomal DNA molecules. Restriction enzymes can be used to insert foreign DNA segments into these circular structures to produce chimeric plasmids. One of the first widely used cloning vectors is the plasmid *pBR322*. This 4363-bp plasmid contains an origin of replication (*ori*) and genes encoding resistance to ampicillin (*amp*) and tetracycline (*tet*).

In this Java applet, you can quickly look up the recognition sequences for any or all of the restriction sites in *pBR322*. Simply click on the name of the restriction site on the plasmid map, and the corresponding cleavage pattern will appear in the box on the right side of the display:

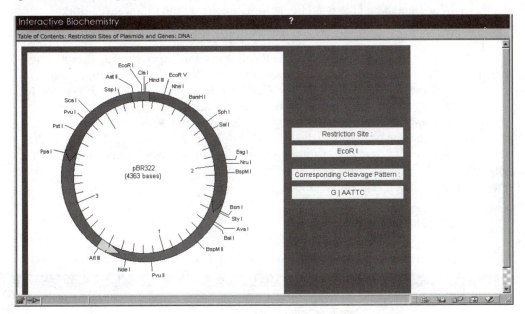

Questions:

1. What is the cleavage pattern for the restriction site *Pst*I?_____

2. What is the cleavage pattern for the restriction site *Bal*I?_____

3. Suppose you are attempting to insert a segment of foreign DNA into the *tet* gene of this plasmid. Explain how you could do this in a directional manner, describing the way in which you would engineer the foreign DNA for insertion as planned._____

_____

Enzyme Kinetics

| See G&G, 2/e |
| p. 434-441 |

For enzymes that obey Michaelis-Menten kinetics, the plot of velocity (v) versus substrate ([S]) is complex. At low [S], the reaction is approximately first-order in [S], and the velocity rises sharply as [S] increases. At high [S], however, the velocity levels off and approaches V_{max} in an asymptotic fashion, which means that the reaction is approaching a zero-order dependence on [S].

The problem that arises from this kinetic behavior is that plots of enzyme kinetic data are not easy to analyze. V_{max} can be estimated only from an extrapolation of the asymptotic approach of v to some limiting value as [S] increases indefinitely, and K_m can be be derived from the value of [S] that gives $v = V_{max}/2$. Several rearrangements of the Michaelis-Menten equation transform it into a straight line equation. The Lineweaver-Burk equation is obtained by taking the inverse of the Michaelis-Menten equation, and the Hanes-Woolf equation is derived by multiplying across the Lineweaver-Burk equation by [S].

In this exercise, Java applets enable the user to examine and manipulate kinetic data, using all three of the graphical presentations described here. More importantly, these plots are interactive. In the applets called "A First Look", the user can explore the effects of varying K_m and V_{max}, either by using the clickable buttons in each applet that change these parameters by fixed amounts or by entering new values for these parameters. In the applets titled "Advanced Modeling", the user can click and drag the plots to vary K_m or V_{max} continuously. By comparing the values for the kinetic parameters with the appearance of the plot, the user can obtain significant insight into the nuances of these kinetic plotting methods.

Michaelis-Menten

Start the applet called "A First Look". The screen will look like this:

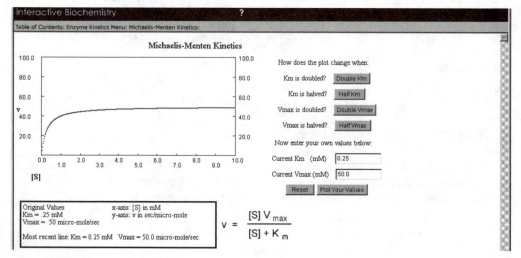

Charles M. Grisham

Notice that the plot is drawn using a K_m of 0.25 mM and V_{max} of 50 micromole/sec. Notice too that there are several clickable buttons on the right side of the display. Starting at the top, click each of these in turn and watch what happens on the plot. When each button is clicked, a new line is drawn on the plot, with the change in parameter (relative to the original value) indicated on the button. Thus, clicking the first button doubles the value of K_m to 0.50 mM, and the new line drawn reflects this change. The second button reduces the K_m by half to 0.125 mM, and so on.

Once you have clicked all four of the buttons, you can continue your exploration by entering your own values for K_m and V_{max} in the boxes indicated. When you do so and click "Plot Your Values", a new curve is drawn. You can plot any number of new curves this way, and you also can erase the added curves and reset the graph to its original appearance with a single curve by clicking "Reset".

Now go back one level in the menu and click on "Advanced Modeling". This will start a Java applet that allows you to continuously vary K_m and/or V_{max} and study the resulting curves as you go.

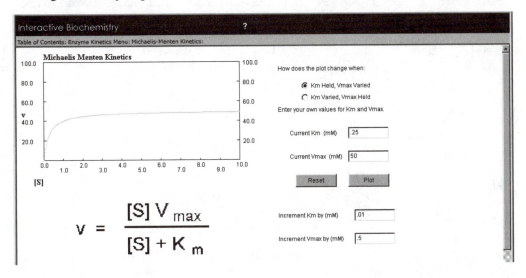

This plot is interactive and live. Simply place the cursor somewhere on the plot itself, and click and drag (up and down or right and left), and the plot will respond by moving on the page. The boxes on the right-hand side of the screen will continuously update the values of K_m and V_{max}.

Note that this applet can be used in two modes which are selected by buttons in the top right of the screen. You can hold K_m constant and vary V_{max}, or you can hold V_{max} constant and vary K_m. In addition, you can enter your own values of K_m and V_{max} in the boxes that display current values of these parameters.

There are two other notable features of this applet: The amounts by which K_m and V_{max} are incremented when the mouse is clicked and dragged over the plot can be varied by inserting appropriate values into the boxes in the lower right of the screen. You should experiment with increment values and observe the results carefully to achieve the most effective control over the parameters displayed and plotted. Also, note that the axes of the plot are automatically rescaled if the parameters you have selected (either by clicking and dragging or by entering values manually) have become inappropriate for the scale of the plot. Keep this rescaling feature in mind as you attempt to interpret the shape and nature of the displayed plot.

Lineweaver-Burk

The Lineweaver-Burk plot is a linear crafting of the Michaelis-Menten equation. This applet operates very similarly to the Michaelis-Menten applet, but the plot offers a different look at the data and at the K_m and V_{max} parameters.

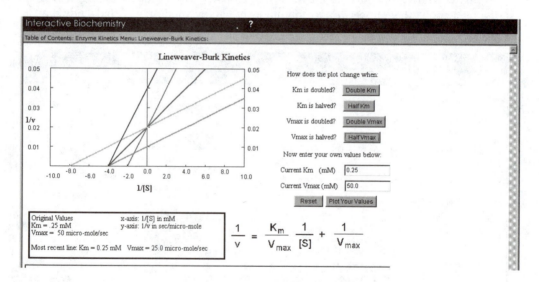

After clicking the four buttons in the upper right of the screen, you may enter your own values for K_m and V_{max} in the boxes that display current values. Click on "Plot Your Values" to display the new line using the parameters you chose. Click on "Reset" to erase the plotted lines and begin again.

Hanes-Woolf

The Hanes-Woolf plot offers yet another linear presentation of enzyme kinetic data. Due to the ways in which errors propagate in the two plots, Hanes-Woolf is superior to Lineweaver-Burk for most plotting purposes. The applet plots [S]/v versus [S] and provides buttons to alter K_m and V_{max} as before:

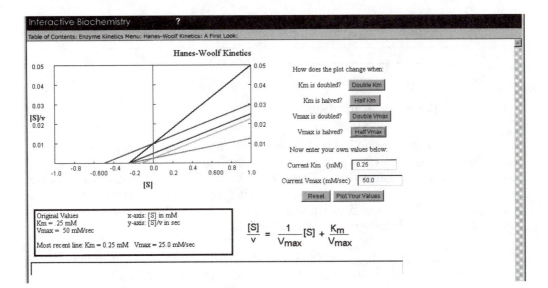

Questions:

1. What happens to the shape and location of the v versus [S] plot when K_m is doubled? When it is halved?_____

_____

2. How does the Lineweaver-Burk plot change when K_m is doubled? When it is halved?

_____

_____

3. How does the Hanes-Woolf plot change when K_m is doubled? When it is halved?____

_____

_____

4. How do each of the plots change when K_m is increased by a factor of 10? Decreased by a factor of 10?_____

_____

5. How do each of these plots change when V_{max} is doubled? When it is halved? When it is increased by a factor of 10? When it is decreased by a factor of 10?_____

_____

_____

6. What happens to each of these plots when K_m is decreased but V_{max} is increased? Experiment with different combinations of these parameters and comment on the results._____

_____

_____

_____

Reversible Inhibition

Many enzymes are subject to reversible inhibition by one or more substances. If the inhibitor and the substrate compete for the same binding site on the enzyme, the inhibition is said to be **competitive.** On the other hand, when the inhibitor binds to a site on the enzyme that is distinct from the substrate site, the inhibition is said to be **noncompetitive**. The choice between these two possibilities often can be made on the basis of kinetic data obtained at several concentrations of inhibitor.

These Java applets begin by displaying the same plots shown in the previous kinetics applets. This time, however, a button in the upper right can be clicked three times to display curves that would arise from three different inhibitor concentrations. When these three plots have been added to the graph, the user can enter new values for the parameters as shown, and the corresponding plots can be drawn.

Competitive Inhibition

In competitive inhibition, the inhibitor increases the K_m for the enzyme by the factor $(1 + [I]/K_I)$. Thus v, the enzyme velocity, is reduced by the inhibitor, but the V_{max} is not affected. Competitive inhibition results in intersection of plotted lines on the y-axis in Lineweaver-Burk plots, but parallel lines in Hanes-Woolf plots.

When you click on the "Competitive Inhibition" selection in the Table of Contents, listed under "Reversible Inhibition", you will be taken to a screen that shows, first, a reaction scheme for competitive inhibition in a Michaelis-Menten model, and, below that, the equations for competitive inhibition in the Michaelis-Menten, Lineweaver-Burk, and Hanes-Woolf formalisms. Each of these has a clickable link that will take you to the Java applet for that case. Click one of these to begin the applet.

_____ Charles M. Grisham

For any of the three formalisms and plots, the interactive exercises are similar, differing only in the form of the graph. When you start the applet, it will look like this:

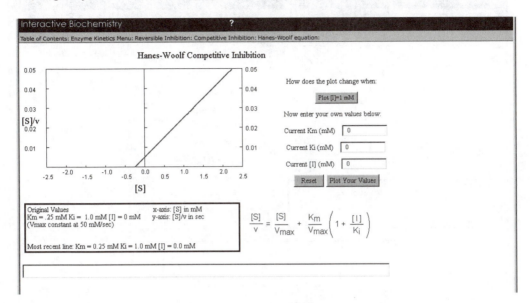

The curve drawn in each case corresponds to a K_m of 0.25 mM and a V_{max} of 50 mM/sec. In the upper right of the screen is a button with which you can plot curves for different concentrations of inhibitor. The assumed value of K_i for these plotted curves will be 1.0 mM. When you begin the applet, the button says:

"Plot [I] = 1 mM"

Click the button and the [I] = 1 mM curve will be drawn. The button also will change to read:

"Plot [I] = 2 mM"

Click the button again and the [I] = 2 mM curve will be drawn. The button also will change to read:

"Plot [I] = 3 mM"

Click the button again and the [I] = 3 mM curve will be drawn. The button also will change to a shadowed image of the previous text, indicating that it is now inactive.

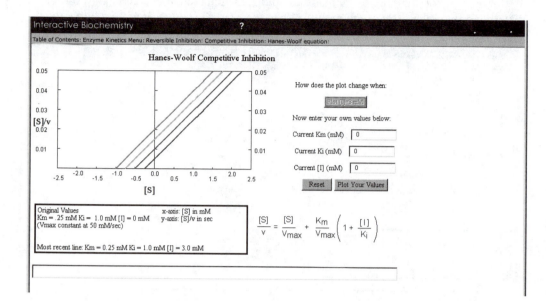

At this point, you can plot additional curves, selecting your choice of values for K_m, K_i, and [I]. This provides a powerful means of quickly analyzing different combinations of these parameters in the formalism of your choice (Michaelis-Menten, Lineweaver-Burk, or Hanes-Woolf).

When you plot additional curves, keep in mind two features of the applet. First, the axes of the graph are "auto-scaling" and will adjust to accommodate your selected values and the resulting plot. Be sure to watch and keep track of the x-axis and y-axis numerical values to compare two or more plots and to assess the adjustments made by the program to accommodate any values you select. Second, when the axes perform autoscaling, you can keep track of previously plotted curves and distinguish them from newly plotted lines, because the colors of previously plotted curves are retained. This feature allows you to plot curves for widely varying values of K_m, K_i, and [I], while keeping track of previously plotted curves.

Noncompetitive Inhibition

In cases of noncompetitive inhibition, the presence of inhibitor increases both the terms in the Michaelis-Menten equation denominator by the factor $(1 + [I]/K_I)$. Thus v, the enzyme velocity, is reduced by the presence of inhibitor, but the V_{max} is not affected. Pure noncompetitive inhibition results in intersections of plotted lines on the x-axis in both Lineweaver-Burk and Hanes-Woolf plots.

When you click on "Noncompetitive Inhibition" in the Table of Contents, listed under "Reversible Inhibition", you will be taken to a screen that shows, first, a reaction

Charles M. Grisham

scheme for noncompetitive inhibition in a Michaelis-Menten model, and, below that, the equations for competitive inhibition in the Michaelis-Menten, Lineweaver-Burk, and Hanes-Woolf formalisms. Each of these has a clickable link that will take you to the Java applet for that case. Click one of these to begin the applet.

For any of the three formalisms and plots, the interactive exercises are similar, differing only in the form of the graph. When you start the applet, it will look like this:

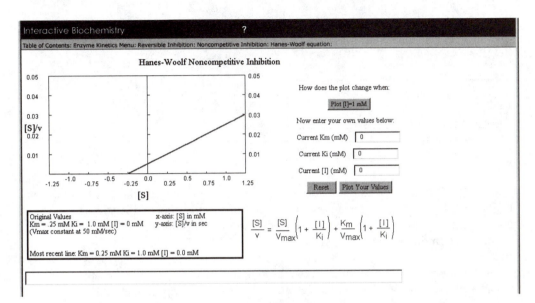

The curve drawn in each case corresponds to a K_m of 0.25 mM and a V_{max} of 50 mM/sec. In the upper right of the screen is a button with which you can plot curves for different concentrations of inhibitor. The assumed value of K_i for these plotted curves will be 1.0 mM. When you begin the applet, the button says:

"Plot [I] = 1 mM"

Click the button and the [I] = 1 mM curve will be drawn. The button also will change to read:

"Plot [I] = 2 mM"

Click the button again and the [I] = 2 mM curve will be drawn. The button also will change to read:

"Plot [I] = 3 mM"

Click the button again and the [I] = 3 mM curve will be drawn. The button also will change to a shadowed image of the previous text, indicating that it is now inactive.

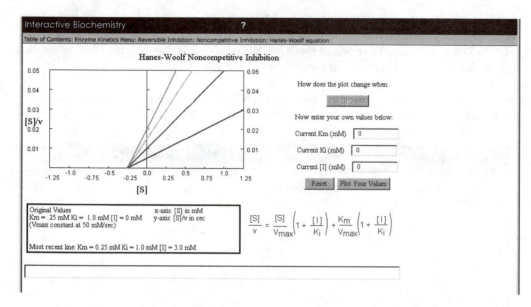

Hanes-Woolf Noncompetitive Inhibition

How does the plot change when:

Plot [I]=3 mM

Now enter your own values below:

Current Km (mM)   0

Current Ki (mM)   0

Current [I] (mM)   0

Reset   Plot Your Values

Original Values
Km = .25 mM Ki = 1.0 mM [I] = 0 mM   x-axis: [S] in mM
(Vmax constant at 50 mM/sec)   y-axis: [S]/v in sec

Most recent line: Km = 0.25 mM Ki = 1.0 mM [I] = 3.0 mM

$$\frac{[S]}{v} = \frac{[S]}{V_{max}}\left(1 + \frac{[I]}{K_i}\right) + \frac{K_m}{V_{max}}\left(1 + \frac{[I]}{K_i}\right)$$

At this point, you can plot additional curves, selecting your choice of values for K_m, K_i, and [I]. This provides a powerful means of quickly analyzing different combinations of these parameters in the formalism of your choice (Michaelis-Menten, Lineweaver-Burk, or Hanes-Woolf).

When you plot additional curves, keep in mind two features of the applet. First, the axes of the graph are "auto-scaling" and will adjust to accommodate your selected values and the resulting plot. Be sure to watch and keep track of the x-axis and y-axis numerical values to compare two or more plots and to assess the adjustments made by the program to accommodate any values you select. Second, when the axes perform autoscaling, you can always keep track of previously plotted curves and distinguish them from newly plotted lines, because the colors of previously plotted curves are retained. This feature allows you to plot curves for widely varying values of K_m, K_i, and [I], while keeping track of previously plotted curves.

Questions:

1. What happens to the shape and location of the plotted curve or line as the inhibitor concentration is increased in each of the three kinetic formalisms (Michaelis-Menten, Lineweaver-Burk, or Hanes-Woolf)?_____

_____

2. What is the relationship between the x-intercept and K_m in the Lineweaver-Burk plot in the absence of any inhibitor?_____

3. What is the relationship between the x-intercept and K_m in the Lineweaver-Burk plot in the presence of a reversible, competitive inhibitor?_____

4. What is the relationship between the x-intercept and K_m in the Hanes-Woolf plot in the presence of a reversible, noncompetitive inhibitor?_____

5. What is the relationship between the y-intercept and V_{max} in the Lineweaver-Burk plot in the absence or presence of a reversible, competitive inhibitor?_____

6. What is the relationship between the y-intercept, V_{max}, and K_m in the Hanes-Woolf plot in the absence and presence of a reversible, noncompetitive inhibitor?_____

7. In any of the Hanes-Woolf plots for an enzyme in the presence of a competitive inhibitor, use the y-intercept to calculate the inhibitor concentration and see if it agrees with the value shown on the screen for that plot._____

8. For any of the Lineweaver-Burk plots for an enzyme in the presence of a competitive inhibitor, use the x-intercept to calculate the inhibitor concentration and see if it agrees with the value shown on the screen for that plot._____

9. For any of the Lineweaver-Burk plots for an enzyme in the presence of a noncompetitive inhibitor, use the y-intercept to calculate the inhibitor concentration and see if it agrees with the value shown on the screen for that plot._____

10. For any of the Michaelis-Menten plots for an enzyme in the presence of a competitive inhibitor, drop a vertical line from the curve at 50% of V_{max} and use the x-intercept to calculate the inhibitor concentration. _____

11. For any of the Hanes-Woolf plots for an enzyme in the presence of a noncompetitive inhibitor, use the y-intercept to calculate the inhibitor concentration and see if it agrees with the value shown on the screen for that plot._____

Enzyme Mechanisms

Introduction

A full appreciation of biochemistry should include an understanding of the mechanisms of enzymatic reactions in living things. The essence of enzyme mechanisms is simple chemistry, and yet the details of mechanisms can be daunting and difficult. This section of *Interactive Biochemistry* provides an interactive way to learn about enzyme mechanisms.

The Table of Contents for the Enzyme Mechanisms section of the CD presents three choices: the Aspartic Protease, Chymotrypsin (a serine protease), and a section on Coenzyme-dependent Reactions. When the cursor is placed on this latter option, a sub-menu is displayed to the right of the main menu to reveal the coenzymes whose reaction mechanisms are included in this section.

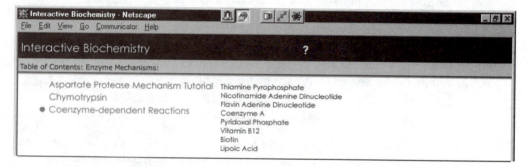

When this section is selected (by clicking on it), a new layer of sub-menus appears for each of the coenzymes listed. For example, the sub-menu for Pyridoxal Phosphate looks like this:

When an enzyme has been selected from these menus and sub-menus, the Java applet will load (this can take several seconds), and the screen will appear something like the figure on the following page.

Charles M. Grisham

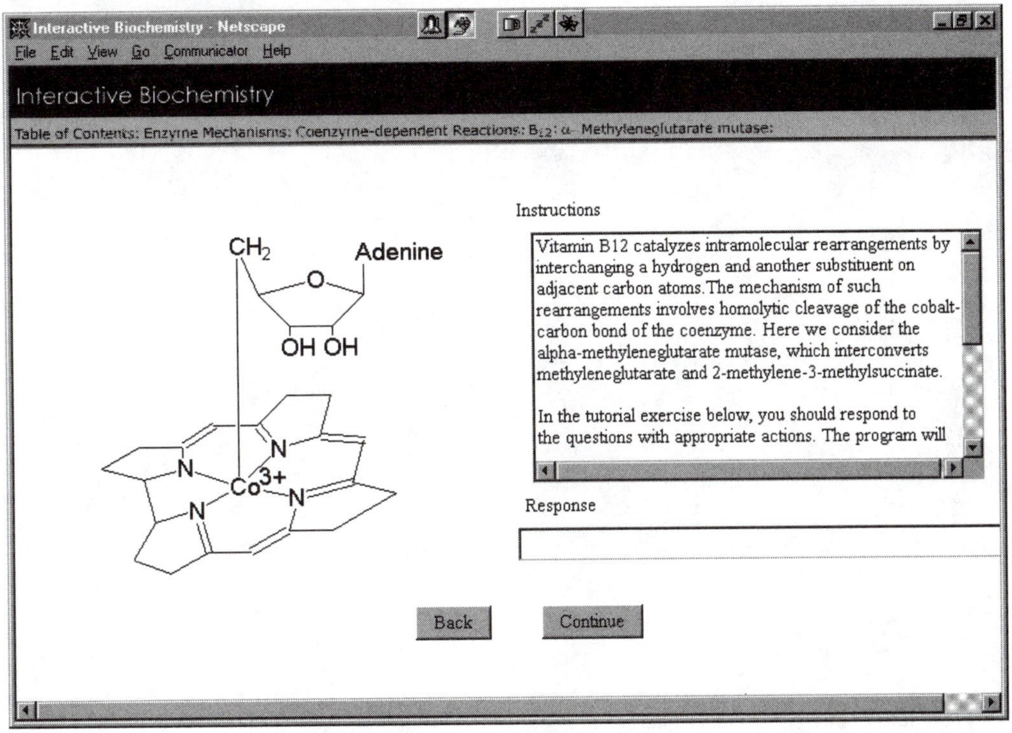

The display includes an area on the upper left for presenting and manipulating structures, an Instructions text box on the upper right, a response box below that, and navigation buttons at the bottom of the screen.

To begin any of these exercises, read the instructions in the text window, study the structure(s) shown at left, and follow the instructions. This will normally require that you click on an atom or bond in the structure(s) shown and then observe the response of the applet. In most cases, when you make the correct response to the problem shown, a blue rectangle will blink on the atom or bond that you clicked, and changes may occur on the structure(s) shown. You also will be prompted to "Click on Continue to Proceed". On the other hand, if the atom or bond you click on is not the correct one, the response box will ask you to try again.

In this way, you will be able to work through each enzyme mechanism, making decisions that illuminate and guide the mechanism itself, and observing the electron movements, bond-making, and bond-breaking, and structural changes that occur in the course of the enzymatic reaction.

The questions posed throughout these exercises, and your answers to them, will emphasize and clarify the mechanistic chemistry that lies within each of these reactions. However, working through one of these mechanisms is just a step in the learning process. The user should conclude each of these exercises by writing out the mechanism by hand on paper. This will help to assure that the insights and details of enzyme mechanisms are grasped and understood fully.

Aspartic Protease
See G&G, 2/e
p. 519-524

Questions:

1. What is the quaternary state of most animal aspartic proteases?_____

2. What is the quaternary state of the aspartic protease from HIV-1?_____

3. What is the catalytic function of the aspartic acid residues in the active site of
   aspartic proteases?_____

4. A water molecule carries out a nucleophilic attack during the aspartic protease
   mechanism. What is the atom that is attacked by the water molecule in this
   mechanism?_____
   .

5. Based on your answer to question 4 above, what would you expect to observe if an
   aspartic protease reaction were carried out in $H_2^{18}O$?_____
   _____

6. Study this mechanism carefully and suggest what the pH-rate profile of an aspartic
   protease should look like. That is, what would be the appearance of a plot of enzyme
   rate versus pH? Why?_____
   _____
   _____

7. What step of this reaction would you expect to be the rate-determining step?_____
   _____

8. How do the drugs known as AIDS protease inhibitors work?_____
   _____

9. Write a chemically correct mechanism for an aspartic protease reaction.

Chymotrypsin

See G&G, 2/e
p. 118, 514-519

Questions:

1. In all serine proteases, including chymotrypsin, three amino acid residues form a "catalytic triad" that mediates the events of catalysis in the active site. What are the three residues that form the catalytic triad? (List the name of the residue and its position in the sequence of the enzyme)._____

2. Serine proteases like chymotrypsin exhibit "burst kinetics". From your study of this mechanism, what do you suppose can be concluded about the mechanism of a serine protease from the observation of burst kinetics?_____

   _____

3. What is the role of the active site aspartic acid in the chymotrypsin reaction?_____

   _____

4. What is the role of the active site histidine in the chymotrypsin reaction?_____

   _____

5. What is the role of the active site serine in the chymotrypsin reaction?_____

   _____

6. A molecule of water carries out a nucleophilic attack in the chymotrypsin reaction. Draw structures to show at what stage of the chymotrypsin reaction this attack occurs and what atom is subject to the nucleophilic attack.

7. What do you suppose is the rate-determining step in the chymotrypsin reaction?_____

   _____

8. Two tetrahedral oxyanions are formed during the course of the chymotrypsin reaction. How are these relatively energetic intermediates stabilized in the active site?_____

   _____

9. What change would you expect in the kinetic properties (K_m and V_{max}) of chymotrypsin if the active site aspartate were changed to an asparagine?_____

   _____

_____ Charles M. Grisham

Coenzyme-Dependent Reactions

Thiamine Pyrophosphate

Thiamine pyrophosphate (TPP) consists of a thiazole ring joined to a substituted pyrimidine by a methylene bridge, with an ethylene-linked pyrophosphate. TPP is a coenzyme involved in reactions of carbohydrate metabolism in which bonds to carbonyl carbons (aldehydes or ketones) are synthesized or cleaved. The four TPP reactions considered in this section share many mechanistic features.

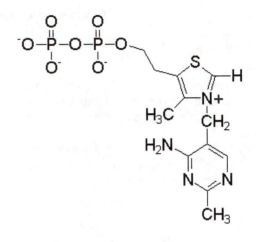

Acetolactate Synthase

See G&G, 2/e
p. 587-588, p. A-17

Questions:

1. What two kinds of reactions are catalyzed by thiamine pyrophosphate?_____
    _____
    _____

2. Which of these (of the answers to question 1) is represented by the acetolactate synthase reaction?_____

3. Draw the two resonance forms of **hydroxyethyl-TPP**.

4. In the acetolactate synthase reaction, the carbanion formed by decarboxylation of pyruvate can react with another molecule of pyruvate to form α-acetolactate. Which atom of the second pyruvate is subject to attack by the carbanion in this reaction?_____
    _____

5. What is the metabolic role or purpose of the α-acetolactate formed in this reaction?___
    _____

6. Is the decarboxylation of pyruvate that initiates this reaction considered an *oxidative* or *nonoxidative* decarboxylation?_____

Phosphoketolase | See G&G, 2/e
p. 587-588, p. A-18

Questions:

1. In the phosphoketolase reaction, fructose-6-P and inorganic phosphate (P_i) react to form acetyl-P and erythrose-4-P. Which of these product molecules contains the phosphorus atom that entered the reaction as inorganic phosphate?_____

2. This reaction, like all thiamine pyrophosphate-catalyzed reactions, depends upon electron withdrawal by an atom on the coenzyme which acts as an electron sink. Draw the structure of TPP showing which atom behaves as the electron sink.

3. Nucleophilic attack by a carbanion on the coenzyme initiates the phosphoketolase reaction. Draw the structure of the coenzyme that shows this carbanion.

4. Which atom of fructose-6-P is attacked by the TPP carbanion?_____

5. In the phosphoketolase reaction mechanism, the hydroxyethyl-TPP intermediate breaks down with production of a water molecule. This is followed by a protonation-deprotonation step, and finally nucleophilic attack by the inorganic phosphate. Draw this sequence of three reactions.

6. The phosphoketolase reaction is referred to as an internal oxidation-reduction reaction, because one atom of an intermediate is oxidized while the other is being reduced. Explain by identifying the two atoms involved and drawing a structure or structures to support your answer._____

_____

Charles M. Grisham

Pyruvate Decarboxylase

See G&G, 2/e
p. 587-588

Questions:

1. Of the two kinds of reactions catalyzed by thiamine pyrophosphate (see question 1 in the acetolactate synthase exercise), which is represented by the pyruvate decarboxylase reaction?_____

2. Is this decarboxylation reaction *oxidative* or *nonoxidative*?_____

3. The cationic imine nitrogen of TPP plays two distinct catalytic roles in the pyruvate decarboxylase and other TPP-catalyzed reactions. What are those?_____
    _____
    _____
    _____

4. In the last step of the pyruvate decarboxylase reaction mechanism, what event must occur before the acetaldehyde product is released from TPP and the enzyme active site?_____

Transketolase

See G&G, 2/e
p. 767

Questions:

1. The transketolase reaction produces the products glyceraldehyde and sedoheptulose-7-P from D-xylulose-5-P and D-ribose-5-P. The reaction involves transfer of a 2-carbon unit. Which of the substrate molecules is the donor of the two-carbon unit?___
    _____

2. Following attack of the TPP anion on the D-xylulose-5-P substrate, a proton abstraction leads to formation of the first product of the reaction. Draw the structure of xylulose-5-P, and show the events of proton abstraction and first product formation.

3. What is the name of the TPP intermediate formed in the reaction described in question 2?_____

4. Nucleophilic attack by the intermediate described in question 3 on D-ribose-5-P initiates the second half of the reaction. Show this attack and the intermediate that is thus formed in this step.

5. Deprotonation of the intermediate described in question 4 leads to formation of the second product of the reaction. Draw structures to show the formation of this second product from the afore-mentioned intermediate.

Nicotinamide Adenine Dinucleotide (NAD)

NAD and NADP play vital roles in a variety of enzyme-catalyzed oxidation-reduction reactions. NAD^+ is a two-electron acceptor in oxidative (catabolic) pathways — and NADPH is a two-electron donor in reductive (biosynthetic) pathways. These reactions involve direct transfer of hydride anion either to NAD^+ or $NADP^+$ or from NADH or NADPH. The enzymes that facilitate such transfers are thus known as dehydrogenases.

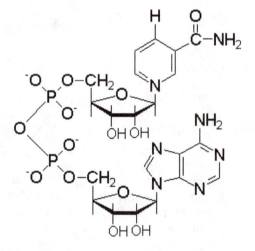

Alcohol Dehydrogenase

See G&G, 2/e
p. 656-657

Questions:

1. In the conversion of ethanol to acetaldehyde, is the required form of the coenzyme NAD^+ or NADH?_____

2. In the conversion of acetaldehyde to ethanol, is the substrate oxidized or reduced?

_____

_____ Charles M. Grisham

3. Is the conversion of ethanol to acetaldehyde a one-electron or two-electron transfer?

_____

3. The two hydrogens of the CH_2 –group of ethanol are chemically equivalent, but distinct in terms of stereochemistry. Yet the carbon at this position is not chiral. What is the term used to describe such an atom?_____

4. What is the oxidation number of the C-1 carbon in ethanol?_____

5. What is the oxidation number of the C-1 carbon in acetaldehyde?_____

Lactate Dehydrogenase | See G&G, 2/e p. 631

Questions:

1. In the conversion of pyruvate to lactate, is the required form of the coenzyme NAD^+ or NADH?_____

2. In the conversion of lactate to pyruvate, is the substrate oxidized or reduced?_____

_____

3. Is the conversion of pyruvate to lactate a one-electron or two-electron transfer?

_____

4. What is the oxidation number of the C-2 carbon in lactate?_____

5. What is the oxidation number of the C-2 carbon in pyruvate?_____

Flavin Adenine Dinucleotide (FAD)

Riboflavin, flavin mononucleotide (FMN)` and flavin adenine dinucleotide (FAD) each contain an isoalloxazine ring system which can exist in any of three redox states, each differing by one electron. Thus FAD and its counterparts are capable of catalyzing *either* one-electron or two-electron transfer reactions. As a result, flavoproteins (with flavins tightly bound) catalyze many different reactions in biological systems and work together with many different electron acceptors and donors, such as NAD, NADP, quinones, and one-electron-transferring proteins such as cytochromes.

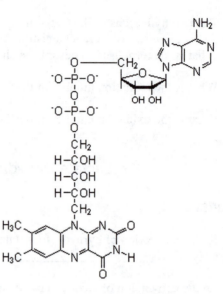

Dihydroorotate Dehydrogenase

See G&G, 2/e p. 915

Questions:

1. In addition to the substrate and FAD, what other electron-acceptor/donor participates in this reaction?_____

2. This reaction involves two hydride transfers. Describe briefly the donor and acceptor for each electron transfer.

	Donor	Acceptor
First electron transfer:	_____	_____
Second electron transfer:	_____	_____

Charles M. Grisham

3. What is the biosynthetic pathway that involves oxidation of dihydroorotaterotate?____

Coenzyme A (CoA)

Coenzyme A consists of 3',5'-ADP joined to 4-phosphopantetheine. The sulfhydryl group can form thioester linkages with various substrates, as in the case of acetyl-CoA` shown at right. The two main functions of coenzyme A are (a) activation of acyl groups for transfer by nucleophilic attack, and (b) activation of the α-hydrogen of the acyl group for abstraction as a proton. Both of these functions are mediated by the reactive sulfhydryl group. CoA thioesters are high-energy molecules, with free energies of hydrolysis in the range of -30 to -35 kJ/mol.

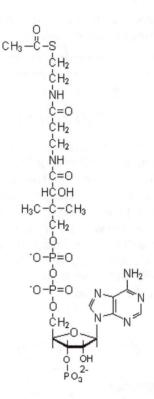

β-Ketothiolase

| See G&G, 2/e p. 788 |

Questions:

1. Which of the two main functions (noted above) of coenzyme A in catalysis are exemplified in the β-ketothiolase reaction?_____

_____

2. Draw portions of the reaction mechanism to illustrate your answer to question 1.

3. Part of the β-ketothiolase reaction depends on the acidic nature of the methyl protons on acetyl-CoA. What would you expect the pK_a of the methyl protons of plain acetate to be by comparison?_____

4. Propose a reasonable chemical explanation for the unusual acidity of the methyl protons of acetyl-CoA._____
_____
_____

Pyridoxal Phosphate (PLP)

Pyridoxal phosphate, shown at right, is the active form of vitamin B_6. It participates in a wide variety of reactions involving amino acids, including transaminations, α- and β-decarboxylations, β- and γ- eliminations, racemizations, and aldol reactions. These reactions include cleavage of any of the bonds to the amino acid alpha carbon, as well as several bonds in the side chain.

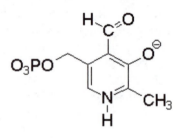

α-Decarboxylation

See G&G, 2/e
p. 595-596

Questions:

1. The remarkably versatile chemistry of pyridoxal phosphate is due to what two chemical properties of this coenzyme?_____
_____
_____
_____

2. How do the properties you described in problem 1 above facilitate the α-decarbox- ylation reaction?_____
_____
_____

Charles M. Grisham

3. Name at least two products of α-decarboxylation reactions in animals, and describe their functions in the organism._____

_____

4. Write a mechanism for the α-decarboxylation of histidine.

β-Decarboxylation

See G&G, 2/e p. 595-596

Questions:

1. Decarboxylation of aspartate at the β-position is more complicated than decarboxylation at the α-position. The chemical essence is that two different carbons on the substrate must be activated. Which are they?_____
_____

2. Following Schiff base formation with pyridoxal phosphate, deprotonation of the substrate initiates this reaction. Which atom of the substrate is deprotonated?_____
_____

3. Once deprotonation has occurred, decarboxylation at the β-position is facilitated by the action of an atom that acts as an electron sink. However, in this case, the electron sink is not the customary one for PLP-catalyzed reactions. What atom acts as the electron sink in the decarboxylation reaction?_____

4. Alton Meister and his coworkers observed in 1964 that aspartate-β-decarboxylase gradually loses activity in the absence of carbonyl compounds or excess PLP. Suggest a reason why this might occur. (See Jencks, W.P., Catalysis in Chemistry and Enzymology, 1969, McGraw-Hill, for an explanation.)_____
_____

γ-Elimination (Methionase)

See G&G, 2/e p. 595-596

Questions:

1. Elimination reactions at the γ-position of amino acids (as in the methionase reaction) require deprotonation at two different positions on the amino acid backbone. Where does the first deprotonation occur?_____

2. Referring to question 1, where does the second deprotonation occur?_____
_____

3. What is the driving force for the elimination at the γ-position in this reaction?_____
_____
_____

4. Two molecules of water are required in the latter stages of this mechanism. Identify the roles of each of these water molecules._____
_____

_____ Charles M. Grisham

Serine Hydroxymethylase (aldol)

See G&G, 2/e
p. 595-596, 882-883

Questions:

1. What are the products of the serine hydroxymethylase reaction?_____
   _____

2. In addition to pyridoxal phosphate, another coenzyme is required for this enzyme
   reaction. What is it?_____

3. Once the PLP-serine Schiff base has formed, this mechanism proceeds with a
   deprotonation, but not at the (usual) α-carbon position. At which position does this
   deprotonation occur?_____

4. Write a simple mechanism for the serine hydroxymethylase reaction.

Racemization

See G&G, 2/e
p. 595-596

Questions:

1. Deprotonation at what position of the substrate follows Schiff base formation in a
   PLP-catalyzed racemization reaction?_____

2. Strictly speaking, "racemization" is an isomerization reaction in which a hydrogen
   atom shifts its stereochemical position at a molecule's *only* chiral center, so as to
   invert configuration at that position. Two amino acids have two chiral centers. What
   are they?_____
   _____

3. Referring to question 2 above, what should be the appropriate term for this same kind
   of reaction (isomerization at a single chiral carbon) in a molecule with two chiral
   centers?_____

Serine Dehydratase

See G&G, 2/e
p. 595-596, 892

Questions:

1. Of the several types of reactions characteristic of PLP (transaminations, racemizations, eliminations, decarboxylations, and aldol reactions), which kind is this?_____

2. As in many other cases of PLP-catalyzed reactions, deprotonation follows Schiff base formation for the serine dehydratase. At which position does this deprotonation occur?_____

3. This reaction involves oxidation at one position of the substrate. Which position?____
_____

4. Referring to question 3, how many electrons are removed from the substrate?_____

5. What is the unstable product that is released from PLP near the conclusion of this reaction mechanism?_____

6. Write the two-step reaction that ensues for the unstable product described in question 5 above.

Transamination

See G&G, 2/e
p. 595-596, 886

Questions:

1. What are the two kinds of substrate that participate in a transamination reaction?_____
_____

2. What is the mechanistic step that follows Schiff base formation at the beginning of the transamination reaction?_____

3. Oxidation and reduction occur in the course of the transamination reaction. Explain._
_____
_____
_____

Charles M. Grisham

4. What is the intermediate state of the coenzyme at the "halfway point" of a transamination reaction?_____

5. "Aldimines" and "ketimines" appear in the course of a transamination reaction. Draw the structures of at least one aldimine and one ketimine from the mechanistic pathway of this reaction.

Vitamin B12

The biologically-active form of vitamin B_{12} consists of a corrin ring with a cobalt ion at its center, with a 5'-deoxyadenosyl group forming one of the axial ligands to the cobalt, and a dimethylbenzimidazole group as the other. Vitamin B_{12} catalyzes intramolecular rearrangements by interchanging a hydrogen and another substituent on adjacent carbon atoms. The mechanism of such rearrangements involves homolytic cleavage of the carbon-cobalt bond of the coenzyme

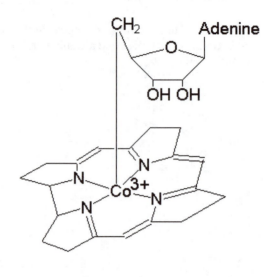

Diol Dehydrase

See G&G, 2/e
p. 597-598, A-19

Questions:

1. All B_{12}-catalyzed reactions begin with the formation of a reactive free radical intermediate. What is the name of the reactive catalytic intermediate formed from the coenzyme?_____

2. What is the name for the type of bond cleavage that forms the coenzymatic free radical intermediate described in question 1 above?_____

3. The diol dehydrase is an intramolecular rearrangement. Draw the structure of 1,2-propanediol, the substrate for this reaction, and indicate the two groups that are interchanged in the enzymatic reaction.

4. When the interchange of the two groups described in problem 3 has occurred, a special kind of diol has been formed. (This diol is unstable and will decompose.) What kind of diol is this?_____

5. What is the name of the product of this reaction?_____

_____ Charles M. Grisham

Ethanolamine Ammonia Lyase

See G&G, 2/e
p. 597-598, A-20

Questions:

1. The ethanolamine ammonia lyase reaction is an intramolecular rearrangement. Draw the structure of ethanolamine, and indicate the two groups that are interchanged in this enzymatic reaction.

2. Of the two interchanging groups you indicated in question 1, which is attacked by the 5'-deoxyadenosyl radical to form the initial free radical on the substrate?_____

_____

3. Draw the structure of the reactive intermediate that is formed by the interchange of groups on the substrate.

4. Write a simple mechanism for the decomposition of the intermediate you drew in question 3.

Glutamate Mutase

See G&G, 2/e
p. 597-598

Questions:

1. Glutamate mutase was the first enzyme to be identified as being B_{12}-dependent. Draw the structure of glutamate, indicating clearly on your drawing which two molecular fragments are interchanged in the glutamate mutase reaction.

2. The product of the glutamate mutase reaction is threo-β-methylaspartate. Draw the structure of this product.

3. To illustrate the changes that occur in this reaction, number the reactant glutamate in question 1 above, beginning with the α-carboxyl carbon and ending with the γ-carboxyl carbon. Then place those same numbers on the product molecule in question 2, to indicate how the order of the carbons has changed after the reaction.

Glycerol Dehydrase

> See G&G, 2/e
> p. 597-598, A-20

Questions:

1. Homolytic cleavage of the Co-carbon bond of vitamin B_{12} yields a reactive free radical that will attack the substrate of this reaction. Draw the structure of the free radical product of vitamin B_{12}.

2. Draw the structure of the glycerol substrate and indicate the two groups that are interchanged in this intramolecular rearrangement.

3. Interchange of the two groups indicated in question 2 yields a geminal diol. Draw the structure of this intermediate in this reaction.

4. Write a reasonable mechanism for the glycerol dehydrase reaction.

Charles M. Grisham

α-Methyleneglutarate Mutase

See G&G, 2/e
p. 597-598, A-19

Questions:

1. Draw the structure of α-methyleneglutarate, number the carbon atoms appropriately, and indicate the two groups that are interchanged in this intramolecular rearrangement.

2. Draw the structure of the product of this reaction, and number the carbons with the same numbers that each had in question 1, in order to illustrate the change that this enzyme accomplishes in the carbon backbone.

Methylmalonate Mutase

See G&G, 2/e
p. 597-598

Questions:

1. This reaction, like all B_{12}-catalyzed intramolecular rearrangements, begins with a homolytic cleavage reaction in the coenzyme. Draw a partial structure of the coenzyme to illustrate this cleavage.

2. Write a reasonable mechanism for the methylmalonate mutase reaction.

Biotin

Biotin acts as a mobile carboxyl group carrier in a variety of enzymatic carboxylation reactions. The biotin ring is tethered to the enzyme by a long molecular chain consisting of five carbons of biotin itself and the side chain of a lysine residue. This chain allows biotin to acquire carboxyl groups at one subsite of the enzyme active site and deliver them to a substrate acceptor at another subsite.

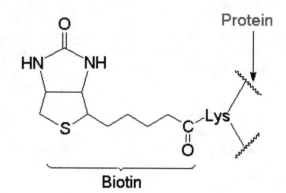

Biotin

Acetyl-CoA Carboxylase

See G&G, 2/e
p. 600, 806

Questions:

1. Acetyl-CoA carboxylase, like other biotin-dependent enzymes, uses bicarbonate as the one-carbon unit. How is this "carboxyl group" activated for transfer?_____
   _____
   _____

2. What is the intermediate that represents the "activated carboxyl group" described in question 1 above?_____

3. Draw the structure of the product of this reaction, and indicate the carbon that was delivered to this product by biotin.

4. Write a mechanism for the formation of N-carboxybiotin in this reaction.

Charles M. Grisham

Beta-Methylcrotonyl-CoA Carboxylase

See G&G, 2/e
p. 600

Questions:

1. Most biotin-catalyzed carboxylation reactions involve delivery of the carboxyl group to a substrate carbanion. In the case of CoA-based substrates, this carbanion is typically alpha to the carbonyl of the thioester. Yet carboxylation in this reaction occurs at a carbon that is gamma to the carbonyl of the thioester. Explain._____

_____

_____

2. Write a mechanism for this reaction that exploits the answer you gave in question 1.

Pyruvate Carboxylase

See G&G, 2/e
p. 600, 746

Questions:

1. Pyruvate carboxylase carries out a carboxylation on the methyl group of pyruvate. Explain how the requisite carbanion is stabilized on pyruvate in this reaction._____

_____

_____

2. Write a mechanism for the pyruvate carboxylase reaction that illustrates the activation of bicarbonate by ATP, the carboxylation of biotin, and the transfer of the one-carbon unit to the methyl carbon of pyruvate.

Lipoic Acid

Lipoic acid exists as a mixture of two forms — the closed form shown at right and an open form that is reduced, with each S existing as an -SH. Complexed to lysine residues on proteins, lipoic acid catalyzes several acyl group transfer reactions. It is found primarily in pyruvate dehydrogenase and α-ketoglutarate dehydrogenase, where it couples acyl-group transfer and electron transfer during oxidation and decarboxylation of α-keto acids.

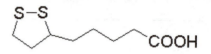

α-Ketoglutarate Dehydrogenase

See G&G, 2/e
p. 601, 652

Questions:

1. α-Ketoglutarate dehydrogenase utilizes five different coenzymes, including lipoic acid. What are the other four?_____

   _____

2. What group is transferred by lipoic acid in this reaction?_____

3. Is the transferred acyl group oxidized or reduced during the course of this reaction?___

   _____

4. What is the function of NAD$^+$ in this reaction?_____

Pyruvate Dehydrogenase

See G&G, 2/e
p. 601, 646-647

Questions:

1. Name the coenzyme that catalyzes decarboxylation of pyruvate in the pyruvate dehydrogenase reaction._____

2. Name the coenzyme that transfers an acetyl group from thiamine pyrophosphate to lipoic acid in this reaction._____

Charles M. Grisham

3. Write a partial reaction mechanism that details the entire cycle of lipoic acid intermediates in this reaction.

4. What is the function of coenzyme A in this reaction?_____
_____

5. What is the function of FAD in this reaction?_____
_____

The Coenzyme-catalyzed Reactions in Metabolism

See G&G, 2/e chapts. 18-28

Introduction

One of the more challenging aspects of biochemistry is the recognition and appreciation of the intrinsic chemistry of metabolic reactions. Much of the challenge centers around the involvement of coenzymes in metabolism. When metabolic pathways require catalytic chemistry that cannot be achieved with the limited arsenal of reactive groups on amino acid side chains, nature invokes the use of coenzymes, organic prosthetic groups that bring novel chemistry to enzyme active sites. Each of these coenzymes possesses well-defined catalytic and chemical capabilities, and the student of biochemistry is well-served if he or she can recognize the chemistry in the reactions involved.

This exercise presents 50 reactions of metabolism (in random order), and asks the user to recognize the chemistry of the reaction and thereby identify the coenzyme required for the reaction. When this Java applet begins, the screen looks like this:

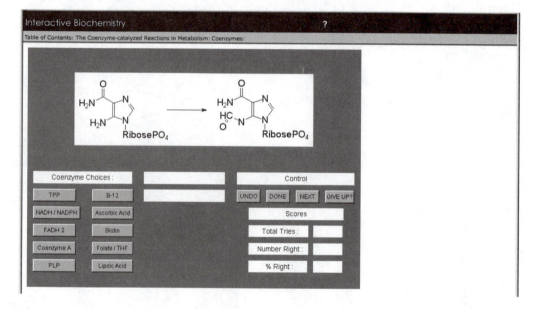

Consider the reaction that appears in the window at the top of the screen, try to recognize the basic chemistry involved in the reaction, and click on the coenzyme (or coenzymes) that you think should be involved in the reaction. When you have selected the coenzymes, click on "Done", and the program will tell you if your choices are correct. If so, you may click on "Next" to proceed to the next, randomly selected reaction. If your answer is not correct, the program will prompt you to try again. You may select a different coenzyme or coenzymes and click "Done" again.

Charles M. Grisham

Questions:

1. What coenzyme is involved in the formation of Schiff base adducts with amino acids and catalyzes a wide variety of reactions with amino acids, including decarboxylations, eliminations, and racemizations?_____

2. What coenzyme is involved in decarboxylations of α-keto acids and the formation and cleavage of α-hydroxyketones?_____

3. What kinds of reactions are catalyzed by coenzyme A?_____

4. What coenzyme is an effective acceptor and donor of one-carbon units for all oxidation levels of carbon except that of CO_2?_____

5. What coenzyme is responsible for transfer of one-carbon units at the oxidation level of CO_2?_____

6. What is the coenzyme required for the following reaction?_____

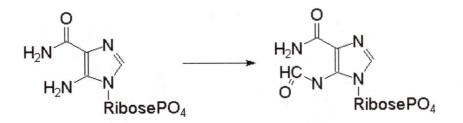

7. What is the coenzyme required for the following reaction?_____

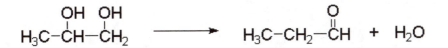

8. What is the coenzyme required for the following reaction?_____

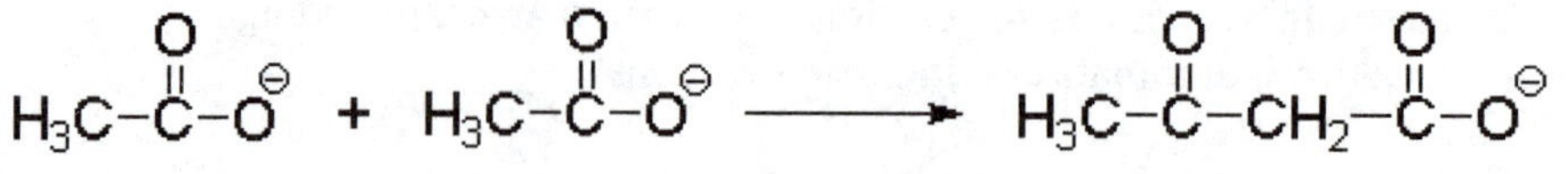

$$\underset{\substack{\text{H}_3\text{C}-\overset{\displaystyle \text{O}}{\overset{\|}{\text{C}}}-\text{O}^{\ominus}}}{} + \underset{\substack{\text{H}_3\text{C}-\overset{\displaystyle \text{O}}{\overset{\|}{\text{C}}}-\text{O}^{\ominus}}}{} \longrightarrow \underset{\substack{\text{H}_3\text{C}-\overset{\displaystyle \text{O}}{\overset{\|}{\text{C}}}-\text{CH}_2-\overset{\displaystyle \text{O}}{\overset{\|}{\text{C}}}-\text{O}^{\ominus}}}{}$$

9. Name the coenzymes that can participate in oxidation/reduction reactions._____

_____

Charles M. Grisham

A Virtual Biochemical World

Introduction

The exercises in this section present three-dimensional views and animations of complex biological processes. They use a language and technology called VRML, short for Virtual Reality Modeling Language. To view these animations, you must have a VRML browser installed on your computer. A VRML browser is a "plug-in" that runs within your web browser. The instructions for finding, downloading, and installing a VRML browser are described in this manual on page xvii (for PC users) and page xxi (for Macintosh users). Follow these instructions before using the exercises and animations.

To use these exercises, simply click on the appropriate entry in the Table of Contents. The VRML browser will open automatically, and the virtual reality scene will be displayed on your screen, together with a text window describing the exercise. One of the valuable and exciting features of VRML technology is that you can manipulate the scene and view it from any point in space. You should feel free to move to different viewing perspectives and reexamine the scene and the animation. Your browser should include a Help option to explain the use of the controls that enable these options.

Actin-Myosin Contraction Model

> See G&G, 2/e
> p. 552-554

Contraction in skeletal muscles involves the sliding of myosin and actin filaments against each other, which results in a net shortening of myofibrils. The model below illustrates how ATP hydrolysis is coupled to this process. The green and orange spheres represent G-actin monomers, arrayed in long filaments. The myosin thick filaments (red) are shown above the actin. For simplicity, only a single myosin head is shown.

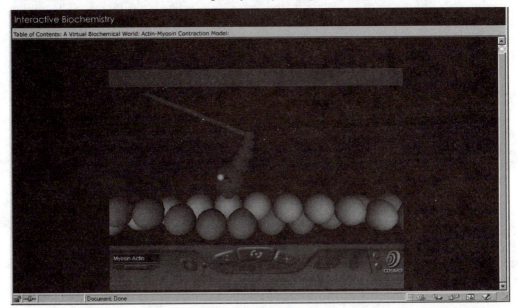

At the beginning of the animation sequence, a previous power stroke has just been completed, and the actin-myosin complex is in the low-energy conformation with the myosin head tightly bound to its site on actin. Subsequent binding and hydrolysis of ATP (adenosine is represented by the yellow sphere, and the triphosphate chain is denoted by blue spheres) causes dissociation of the myosin head from the actin thin filament and also causes the myosin head to shift back to its high-energy conformation with the head's long axis nearly perpendicular to the long axis of the thick filaments. This is the normal resting state of skeletal muscle. When the signal to contract is presented, the myosin heads move out from the thick filaments to bind to sites on the actin thin filaments. Binding to actin stimulates release of phosphate (small blue sphere), and this is followed by the crucial conformation change in the myosin heads — the power stroke — and ADP dissociation. In this step, the thick filaments move along the thin filaments as the myosin heads relax to a lower-energy conformation. In the power stroke, the myosin heads tilt by approximately 45 degrees and the conformational energy of the myosin heads is lowered by about 29 kJ/mol. This moves the thick filament approximately 10 nm along the thin filament.

ATP Synthase Motor

See G&G, 2/e
p. 694-696

The mitochondrial complex that carries out ATP synthesis is the ATP synthase. It consists of two principal complexes, the F_1 unit (with alpha, beta, gamma, delta, and epsilon subunits), which catalyzes ATP hydrolysis, and the F_0 unit (with a, b, and c subunits), which spans the membrane and permits transmembrane proton movement. In the model shown here, the a subunit (brown) is a stator (the stationary component of a motor) to which b (orange) and c (bright blue) subunits bind. Two b subunits extend up to contact and secure the alpha and beta subunits (blue and green), and delta subunits (purple) of F_1, and a ring of 9 c subunits (the "rotor") can rotate in the bilayer. The ring of c subunits is coupled to the gamma and epsilon subunits of F_1, so that rotation of the ring of c subunits causes the γ subunit to rotate inside the ring of alpha and beta subunits in F_1.

This model shows how proton transport can be coupled to ATP synthesis. According to Paul Boyer's binding change mechanism, there are three possible states for each of the beta subunits in F_1 — the ADP/P_i bound state, the ATP bound state, and the free state (no substrates or products bound at the active site). Conformation changes induced by the rotating gamma subunit change the state of each beta subunit sequentially from ADP/P_i-bound to ATP-bound to free.

How does the c rotor rotate to drive ATP synthesis? The c rotor subunits each carry an essential residue, Asp[61]. Rotation of the c rotor relative to the stator may depend upon neutralization of the negative charge on each c subunit Asp[61] as the rotor turns. Protons taken up from the cytosol by one of the proton access channels in the a subunit could protonate an Asp[61] and then ride the rotor until they reach the other proton access channel on a, from which they would be released into the matrix of the mitochondria.

Charles M. Grisham

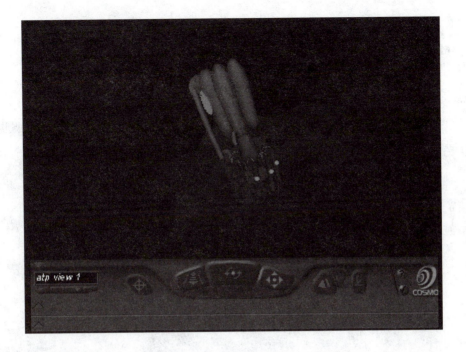

Try viewing this model from different perspectives to appreciate the details of the rotation of the motor and the movement of protons across the membrane.

A Proton Gradient Drives Flagellar Rotation

See G&G, 2/e
p. 561-563

Bacteria swim and move by rotating their flagella. The rotation of the long flagellar filament is the result of rotation of motor protein complexes in the bacterial plasma membrane. The direction of rotation of these filaments affects the movements of the cell. When the half-dozen filaments on the surface of the bacterial cell rotate in a counterclockwise direction, they twist and bundle together and rotate in a concerted fashion, propelling the cell through the medium. (On the other hand, clockwise-rotating flagella cannot bundle together, and under such conditions the cell merely tumbles and moves erratically.)

The rotations of bacterial flagellar filaments are the result of the rotation of motor protein complexes in the bacterial plasma membrane. The flagellar motor consists of at least two rings, including the S ring (pale green) and the M ring (blue center ring), with diameters of about 25 nm assembled around and connected rigidly to a rod attached in turn to the helical filament. The rings are surrounded by a circular array of membrane proteins, including the motB proteins (magenta) on the edge of the M ring and the motA proteins (gray) located in the membrane and facing the M ring.

A proton gradient drives the flagellar motor. The concentration of protons is typically higher outside the cell than inside, resulting in a thermodynamic tendency for protons to move into the cell. The motA and motB proteins together form a proton channel through which proton movement is coupled to the motion of the motor disk. Proton movement into the cell through this channel drives the rotation of the flagellar motor.

Howard Berg and his coworkers have proposed the model shown here. The motA proteins each possess a pair of "half-channels", with one half-channel facing the inside of the cell and the other facing the outside. The motB proteins, on the other hand, possess proton exchange sites — probably aspartate or glutamate carboxyls or imidazole moieties on histidine residues. The outside edges of the motA proteins cannot move past a proton-exchange site on motB when that site has a proton bound, and the center of the motA protein cannot move past an exchange site when that site is empty. As shown in the model, these constraints lead to coupling of proton translocation and flagellar motor rotation. Protons (white spheres) enter the membrane from outside the cell through one half-channel on motA, are then transferred to motB, and then back to inward-facing half-channels on the motA proteins, from which they are released into the cell. The motA proteins are assumed to be elastically linked to the cell wall, such that each movement of a proton through the membrane causes the flagellar wheel to ratchet, with a complete rotation of the wheel requiring the transport of many protons. The model shown here is simplified, but a variety of evidence indicates that 800 to 1,200 protons flow through the motor protein complex during a single rotation of the flagellar filament.

Charles M. Grisham

Peptide Synthesis on the Ribosome

See G&G, 2/e
p. 1101-1102

The synthesis of proteins occurs on ribosomes, which act as docking stations for molecules of tRNA (transfer ribonucleic acid). Only two tRNA molecules are part of the ribosome:mRNA complex at any moment. Each lies in a distinct site. The A, or acceptor site, is the attachment site for an incoming aminoacyl-tRNA. The P, or peptidyl, site is occupied by peptidyl-tRNA, the tRNA carrying the growing polypeptide chain. The elongation reaction transfers the peptide chain from the peptidyl-tRNA in the P site to the aminoacyl-tRNA in the A site. This transfer occurs through covalent attachment of the peptidyl alpha carboxyl to the alpha-amino group of the aminoacyl-tRNA, forming a new peptide bond. The new, longer peptidyl-tRNA now moves from the A site into the P site as the ribosome moves one codon further along the mRNA. The A site, left vacant by this translocation, can accept the next incoming aminoacyl-tRNA.

Termination of this process occurs when the ribosome reaches a "stop" codon on the mRNA. At this point, the polypeptide chain is released, and the ribosomal subunits dissociate from the mRNA.

Protein synthesis proceeds rapidly. In vigorously growing bacteria, about 20 amino acid residues are added to a growing polypeptide chain each second. An average protein with 300 amino acids is thus synthesized in only 15 seconds. Eukaryotic protein synthesis is only about 10% as fast, however.

Calcium-induced Calcium Release

See G&G, 2/e
p. S-16 – S-17

Calcium ion is an important intracellular signal and stimulates a variety of cellular processes. Cytoplasmic calcium can be increased in two ways. Cyclic AMP (cAMP) can activate the opening of plasma membrane calcium channels, allowing extracellular calcium to stream in. On the other hand, cells also contain intracellular reservoirs of calcium, within the endoplasmic reticulum (ER) and calciosomes, small membrane vesicles that are similar in some ways to muscle sarcoplasmic reticulum. These special intracellular calcium stores are not released by cAMP. They respond to inositol trisphosphate (inositol-1,4,5-P_3 or IP_3), a second messenger derived from phosphatidylinositol.

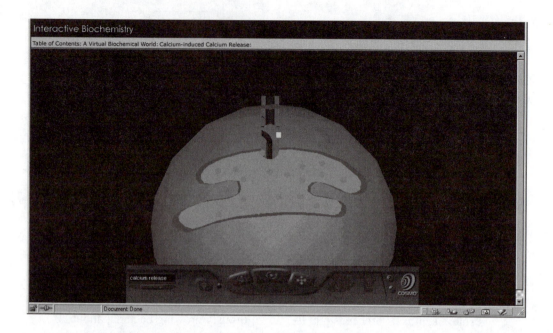

Much of the calcium entering the cytoplasm (red in the model) by action of IP_3 appears to come from two sources: (a) parts of the endoplasmic reticulum (yellow in the model) that are closely associated with the plasma membrane, and (b) the extracellular environment. As shown in the model, calcium release is a two-step process. IP_3 (yellow square) binding to receptors on the ER membrane opens calcium channels (blue), releasing calcium (green spheres) from the ER. Flow of calcium through these channels induces a conformational change that opens plasma membrane calcium channels (purple). This latter channel-opening event is also mediated by inositol-1,3,4,5-P_4.

Charles M. Grisham

Treadmilling by Microtubules

See G&G, 2/e
p. 534-536

Cells have intricate "skeletons" — called cytoskeletons — that give them shape and that carry out a variety of essential cellular processes and movements. Microtubules made from tubulin proteins are one component of the cytoskeleton. Tubulin is a dimeric protein composed of two similar 55-kD subunits known as alpha-tubulin and beta-tubulin. Tubulin dimers polymerize to form microtubules, helical structures with 13 tubulin monomer "residues" per turn. Microtubules grown *in vitro* are dynamic structures that are constantly being assembled and disassembled. Moreover, because tubulin dimers in a microtubule are all oriented similarly, microtubules are polar structures. The end of the microtubule at which growth occurs is the plus end, and the other is the minus end. Microtubules *in vitro* carry out a GTP-dependent process called treadmilling, in which tubulin dimers are added to the plus end at about the same rate at which dimers are removed from the minus end. The animation in this model portrays the treadmilling process, with the growing end (the plus end) at the top and the minus end at the bottom.

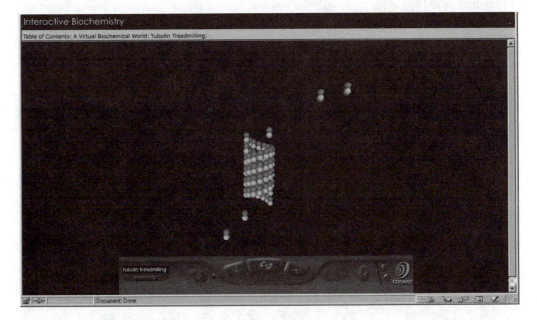

The Action of Flippases

See G&G, 2/e
p. 268

Phospholipids often are distributed asymmetrically across many membranes. In the erythrocyte, phosphatidylcholine comprises about 30% of the total phospholipid in the membrane. Of this amount, 76% is found in the outer monolayer and 24% is found in the inner monolayer. How are transverse lipid assymetries like this created and maintained in cell membranes? From a thermodynamic perspective, these asymmetries could only occur by virtue of asymmetric syntheses of the bilayer itself or by energy-dependent asymmetric transport mechanisms. Asymmetric synthesis of phospholipids does indeed occur, but proteins that can "flip" phospholipids from one side of a bilayer to the other also have been identified in several tissues. Called "flippases", these proteins reduce the half-time for phospholipid movement across a membrane from 10 days or more to a few minutes or less. Some of these systems may operate passively, with no apparent input of energy, but passive transport alone cannot establish or maintain assymetric transverse lipid distributions. However, rapid phospholipid movement from one monolayer to the other occurs in an ATP-dependent manner in erythrocytes. Energy-dependent lipid flippase activity may be responsible for the creation and maintenance of transverse lipid asymmetries.

In this model, phospholipid molecules are shown approaching a flippase protein in one monolayer of the bilayer, binding to the protein, then flipping to the other monolayer and diffusing away.

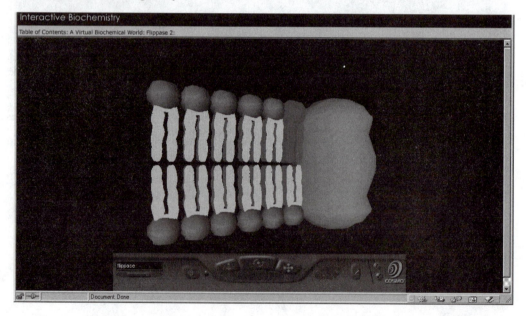

Charles M. Grisham

The Metabolic Map as a Modeling Database

See G&G, 2/e
chapts. 18-28

Metabolism, the sum of the chemical changes that convert nutrients into energy and the chemically complex finished products of cells, consists of literally hundreds of enzymatic reactions organized into hundreds of pathways. These pathways proceed in a stepwise fashion, transforming substrates into end products through many specific chemical **intermediates**. Metabolism is sometimes referred to as **intermediary metabolism** to reflect this aspect of the process. Metabolic maps portray virtually all of the principal reactions of the intermediary metabolism of carbohydrates, lipids, amino acids, nucleotides, and their derivatives. These maps are very complex at first glance and seem to be impossible to learn easily. Despite their appearance, these maps become easy to follow once the major metabolic routes are known and their functions are understood.

Despite their elegance, metabolic maps in their usual form do not always provide sufficient, conveniently available information. For example, typical metabolic maps list the names of metabolites and the EC (Enzyme Commission) numbers for the enzymes that interconvert them, without providing structures of the metabolites or the full names of the enzymes. Moreover, whether or not two-dimensional structures of metabolites can be found in a map, the three-dimensional structures of metabolites often are not easily grasped from such a map. Similarly, users of metabolic maps are frequently curious about the structures of the enzymes involved in particular reactions, and there are few if any resources that describe the known structural features of metabolic enzymes in the context of a metabolic map.

The metabolic map database on this CD-ROM addresses these issues in an accessible and easily usable software environment. When the program is started, the user is presented with an image* of a full metabolic map, where the metabolites are represented by dots and the lines between the dots represent the reactions that interconvert them. (See next page.) The map is color-coded to indicate the particular metabolic purposes and functions of each part of the map. For example, carbohydrate metabolism is shown in green, whereas amino acid metabolism is displayed in red.

Helpful Hint!
This Metabolic Database applet functions by searching for data on a server computer at a site distant from your computer. Thus, to use this database you must have a live network connection to the Internet, either via a direct network connection or a suitable modem/dialup link.

---

*Adapted from Alberts, B., et al., 1989, Molecular Biology of the Cell, 2nd edition, New York: Garland Publishing Co.

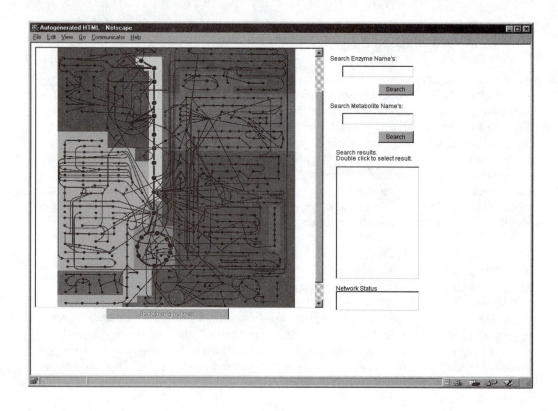

Helpful Hint!

This program enables the user to save files to disk, a procedure that is not usually possible in a web browser. To enable saving files to disk, the user must grant permission. When the Metabolic Database is started, a window will appear that asks the user to grant permission to override the usual browser security procedures. Click the "Grant" button to proceed with the program. This will enable the program to store web pages of data retrieved from the database on your computer's disk drives.

Charles M. Grisham

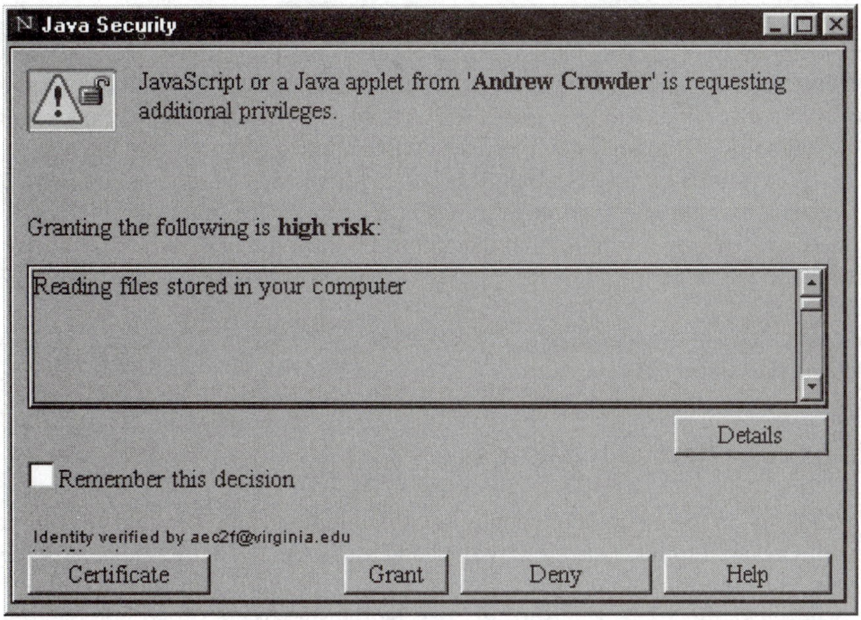

When the mouse is used to scroll the cursor across the map, a yellow label appears to the right of the cursor, displaying the name of the particular portion of the map where the cursor is presently located. If the mouse is clicked while the cursor is in one of these regions, the display will change to a much more detailed version of that portion of the map:

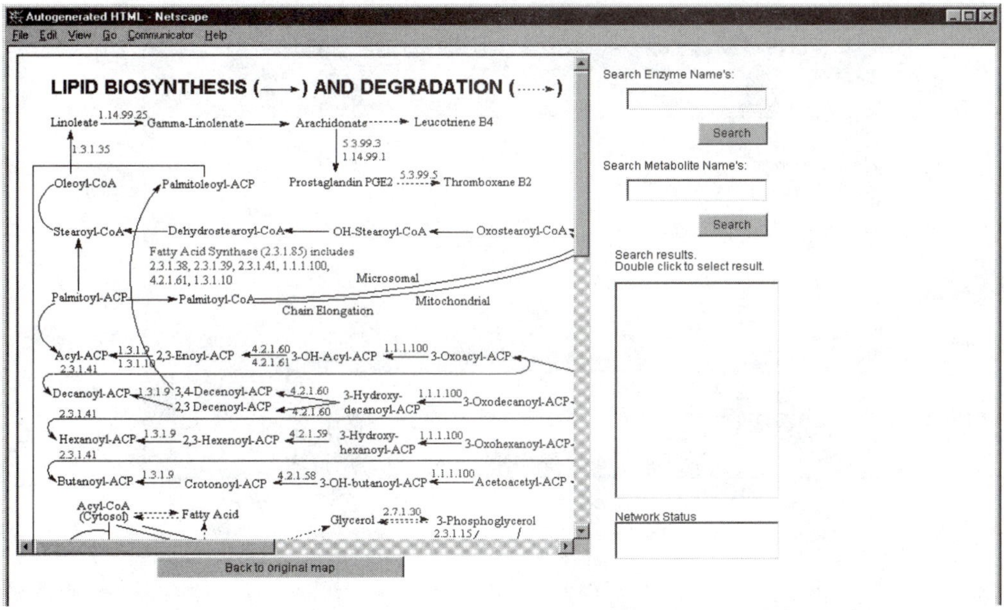

These more detailed maps* display the names of the metabolites in each pathway, together with the EC (Enzyme Commission) numbers for the enzymes for each reaction, in the conventional manner. However, there are several differences in these maps compared to conventional ones. First of all, each of the metabolites and enzymes in the map is "clickable". Clicking on any of these will initiate a search of the metabolic database, as explained below. Second, text boxes to the right of the metabolic map enable the user to enter names of metabolites and enzymes for searches in the database. Either method (clicking on an item on the maps or entering a name in the text boxes) will yield results in the same format.

*Adapted from map designed by and courtesy of D.E. Nicholson, University of Leeds, U.K., and the Sigma Chemical Co., 20th edition, 1997.

Clicking on an Item in the Maps

Clicking on the name of a metabolite or the number (in red) of an enzyme in one of the pathways will initiate a search by the database applet. To use this database effectively, it is important to understand what happens when this occurs. When the search is initiated, the applet sends a call over the Internet to the server computer at Saunders College Publishing that maintains the database files. The server searches the files for information on the metabolite or enzyme requested, and sends that information back over the Internet to your computer, where the data obtained is stored in a file that must be saved to a folder (directory) on a drive on your computer (for example, your hard disk or a floppy disk drive). You will be prompted to select a folder (directory) in which to store this file:

Charles M. Grisham

Select a suitable folder (directory) and file name, and click on "Save". The data will be stored as an HTML page file in the folder you have chosen, and then it also will be displayed on your screen. A search for a metabolite results in a screen like this:

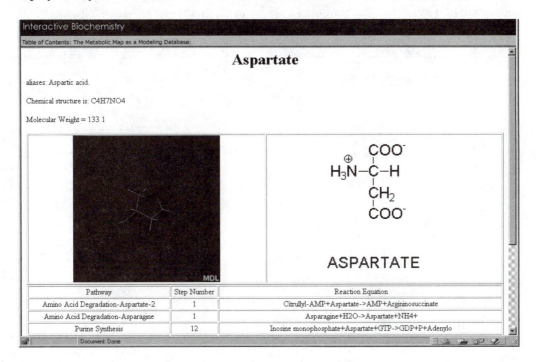

Shown at the top of the metabolite data page are alternative names (aliases) for the metabolite, its atomic composition, and its molecular weight. Below these items, you will see two image areas, side by side. On the left is a window that, using Chime, shows the three-dimensional structure of the metabolite for which you searched, and on the right is a two-dimensional drawing of the structure. The Chime structure, as usual, can be rotated and manipulated in a variety of ways. Take the opportunity to change the display, rotate or zoom the structure according to the instructions provided in the Protein Structure Tutorial Exercises section of this manual. Finally, below the Chime and 2D structures, you will find a table of all the metabolic pathways in which the metabolite participates. The table shows the reactions, the pathway names, and the number of the step in that pathway which involves the metabolite of interest.

Entering a Name for a Search

Searches for metabolites and enzymes also may be conducted by entering a name, or portion of a name, of a metabolite or enzyme in the boxes on the right of the screen:

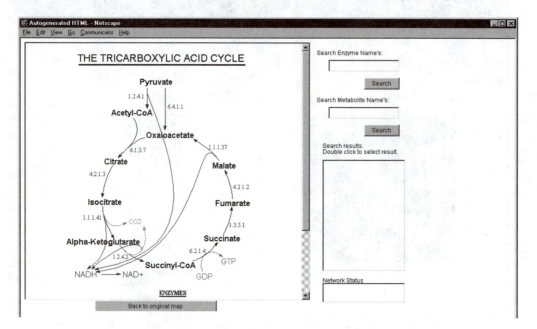

When you enter a name in either of the boxes at the top right of the screen and click on "Search", the applet contacts the server computer over the Internet, and returns with a list of items that fit the search criterion you have entered. Double click on an item from the search list and the applet will proceed as before, opening a "Save as" window and prompting you to select a folder (directory) and file name in which to save the retrieved information as an HTML file.

Charles M. Grisham